Bedouins
by the Lake

Bedouins
by the Lake

Environment, Change, and
Sustainability in Southern Egypt

Ahmed Belal, John Briggs,
Joanne Sharp, Irina Springuel

The American University in Cairo Press
Cairo   New York

Designed by AWH Hughes. Maps and illustrations by Mike Shand.

Dar el Kutub No. 4404/08
ISBN 978 977 416 198 8

Dar el Kutub Cataloging-in-Publication Data

Belal, Ahmed
      Bedouins by the Lake: Environment, Change, and Sustainability in Southern
    Egypt / Ahmed Belal.—Cairo: The American University in Cairo Press, 2008
        p.   cm.
        ISBN 977 416 198 X
        1. Bedouins—Egypt
        305.90691

1 2 3 4 5 6 7 8   14 13 12 11 10 09

Printed in Egypt

Dedication

On 4 January 1999, when returning to Aswan from fieldwork in Wadi Allaqi, one of our field cars was involved in a fatal accident in the desert to the south of Wadi Umm Hibal, about 65 kilometers south of Aswan. Four of our colleagues and friends were killed in that accident. This book is dedicated to their memory.

Mohamed Sogheir
Douglas Ball
Ahmed Hamza
Hashim Rashid

Contents

Acknowledgments

In a twenty-year research project, there are inevitably a large number of people to be thanked for their support and inputs of varying kinds. Every one of the seventy-seven names that appear in Appendix 1 has contributed to the work of the Allaqi Project to a greater or lesser extent. Without all their work and inputs, this book would never have seen the light of day. The four of us have had the privilege of drawing together all the work of the project for this book, but it is the work of all the team that is represented here in the pages that follow. However, there are some people whose contributions stand out and they must be acknowledged separately. Magdi Ali, Mohamed Gabr, Arafa Hamed, Abd El-Moneim Mekki, Usama Radwan, and Abd El-Samie Shaheen were there right at the start of the Allaqi Project in 1987, and have been involved continuously ever since. There are others who may not have been involved continuously for the twenty years of the project, but who nevertheless have made outstanding contributions in terms of time, effort and commitment. Particular mention needs to be made of Gordon Dickinson, Samir Ghabbour, Nabila Hamed, Haythem Ibrahim, Brenda Leith, Sayed Nour El-Din Moalla, Kevin Murphy, Ahmed El-Otify, Ian Pulford, Magdy Radi, Tarek Radwan, Hassan Sogheir, Jacqueline Solway, Wafaa Sorour, Mustafa Taher, and Hoda Yacoub. Our sincere thanks for all your inputs and efforts to the project over many years. This is not to diminish the contributions of all the other people listed in Appendix 1—far from it, and so a most heartfelt thanks to everyone involved.

As a team, we do owe a huge debt of gratitude to Professor Mohamed Kassas, Emeritus Professor of Botany at Cairo University. Professor Kassas has been an enormous source of support for the all of us involved in the Allaqi Project. He has been a tireless supporter of our work, and has unfailingly provided us with encouragement and direction when we needed it. Many thanks indeed. In addition, we received considerable encouragement from Dr. Mostafa Tolba over the lifetime of the project, but particularly so in the crucial early stages of our work, and we are very grateful for that. We are also greatly indebted to Peter Snelson, who worked tirelessly to create our administrative structures in the early years of the project. It is a tribute to his sterling efforts that these structures have served us well over many years subsequently. Once again, many thanks.

We had the considerable advantage in our work of having the support of the British Council who supported the project through the Higher Education Links scheme from 1987 to 2004, at which point DFID, as the funders of the scheme, terminated the entire HEL programme globally, a decision that many in the Global South still find hard to understand. We have been particularly well served by a succession of committed British Council officers who supported us and encouraged us in our collaborative endeavors. We are extremely grateful to these officers, namely Peter Llewellyn, Julian Edwards, Anna Baker, Martin Daltry and Mike Coney, and to Robin Sowden who helped to set the whole project on its way in 1987.

We gratefully acknowledge UNESCO for their recognition of our research in Wadi Allaqi and for their designation of the Wadi Allaqi Biosphere Reserve within the global UNESCO MAB program. We are particularly grateful to Peter Dogse, from UNESCO in Paris, for his continuous interest in our work, and to the UNESCO Cairo office, with special thanks to Abdin Salih, Adnan Shihab-Eldin and Mohamed El-Deek.

We were fortunate to be supported financially over the years by a range of funders. We are particularly grateful for the financial support received from the British Council, the UK Overseas Development Administration (ODA), the United Nations Environmental Programme (UNEP), the United Nations Educational, Scientific and Educational Organization (UNESCO), the International Development Research Centre (Canada), the Gilchrist Educational Trust, and the UK Department for International Development (DFID). We are also very grateful for the support received from within Egypt, and particularly the Supreme Council of Universities, the Ministry of International Co-operation and the Ministry of Agriculture.

Producing this book has been quite a task for the four of us, but we have been extremely well-supported by Caryll Faraldi who has done an immense amount of proof-reading, not just of this manuscript, but of many others over many years of support for the Allaqi Project; by Mike Shand of the Department of Geographical and Earth Sciences of the University of Glasgow who produced all the maps and diagrams in this book, and, like Caryll Faraldi, has provided many years of support for the

Allaqi Project in producing high-quality maps and diagrams for us over many years; and by Chip Rossetti, and subsequently the editorial team, at the American University in Cairo Press who have supported this book project all the way through.

Last, but not least, we want to thank the Bedouin of Wadi Allaqi who have worked with us for more than twenty years on this project, and especially Ali Suleiman, Ali Salih, Mohamed Salih, Ali Hadaya, and the family of the late Jar al-Nabi. They have always been unfailingly hospitable and it has been our pleasure and privilege to work with them as friends and colleagues.

Preface

This book is an account of more than twenty years' continuous research conducted in Wadi Allaqi in the extreme south of Egypt to the east of Lake Nasser. Irina Springuel had started to work on the botany of the Eastern Desert during the late 1970s and the early part of the 1980s, building on the earlier work of Professor Mohamad Kassas of the University of Cairo. By 1987, Dr. Springuel recognized that there were wider issues and challenges at play, however, and saw that the flora of the Eastern Desert, and especially the vegetation that had been modified along the shores of the new Lake Nasser, had potentials for development within local Bedouin communities. There were also other groups from outside the region that equally saw potentials for development, but there were risks involved here as the fragile resource base could be irrevocably destroyed with some of these large-scale development ideas, some of which included extensive irrigation scheme proposals.

Having convinced the then-dean of the Faculty of Science of the then-Assiut University (Aswan Branch), Professor Ahmed Belal, that there were some pressing environmental and development issues at stake, contact was made with the British Council in Cairo to sound out whether there were any researchers in the United Kingdom who might have an interest in pursuing these research interests in collaboration with Egyptian colleagues in Aswan. John Briggs and Gordon Dickinson from the University of Glasgow in Scotland visited Aswan in February 1987, supported by the British Council, and from these modest beginnings the Allaqi Project was born. None of us in 1987 could have imagined the extent to which the project would grow and how it would become a major part of our lives over the following twenty years.

From the outset, we established some important principles underlying the Allaqi Project. First of all, the project needed to be interdisciplinary.

We quickly recognized that no one discipline could provide the answers to and understandings of the complexities and interrelationships of the ecological, environmental, socioeconomic, and developmental challenges of the Wadi Allaqi area. It was important to involve a range of research skills and traditions and there was a real excitement for all of us to be working with researchers from very different traditions—biologists, chemists, social scientists, geographers, ecologists, physicists—but all of us with one shared overriding interest in the environment and development concerns of Wadi Allaqi. Secondly, the project was to be truly collaborative, with shared responsibilities among all contributors, and these also included the Bedouin communities of Wadi Allaqi, many of whom became good friends with researchers. Third, we were determined that the project would also provide an opportunity for the research training of young staff, capacity building for the future of Egyptian science. This included research visits for such staff to Glasgow to receive training that could not at that time be provided in Aswan, and the production of large numbers of postgraduate dissertations and theses as a product of the whole research training process. Finally, and perhaps most importantly, we were committed to working with the Bedouin communities of Wadi Allaqi and involving them in the research process.

The Allaqi Project, however, was more than just a research project. It was able to use its results and findings to achieve some concrete outcomes. We were able to persuade the authorities to designate Wadi Allaqi initially as a protected area, and subsequently as a United Nations Educational, Scientific and Cultural Organization (UNESCO) Biosphere Reserve, and therefore to protect the environment from some of the potential excesses of developers from outside the area. This also ensured that the livelihoods of the Bedouin communities of the area were assured and that they would not have to be 'moved on.' The Aswan governorate was convinced to locate a basic health clinic, once it was assured that there was a sufficiently large 'resident' population in the area. In the early 2000s, a literacy program was established to provide basic reading and writing skills to Bedouin children, something that was subsequently adopted by some of the adults as well.

Perhaps the biggest surprise for us, though, is the extent to which the number of contributors grew during the lifetime of the Wadi Allaqi project. Back in 1987, the project started with a handful of keen and dedicated researchers, but, twenty years later, some seventy-seven people have made significant contributions to the Allaqi Project; their names are listed in Appendix 1, and their associated publications in Appendices 2 and 3 at the back of this book. Although the four of us have taken the responsibility for writing *Bedouins by the Lake*, based on this work, it could not have been written without the enormous contributions made by so many of the Allaqi Project team, a large, multi-disciplinary team of Egyptian, British, Canadian, and Dutch scientists and researchers. This has been a genuine team effort. In each chapter, we have referenced the key Working Papers, and journal and book publications. Our task as the four editors has been to summarize and assimilate these publications to produce what we hope has turned out as a coherent story, but a story that could not have been told without the inputs and huge efforts of the entire Allaqi Project team.

Introduction

This book is about how various groups of people, or stakeholders, have responded to the challenges and opportunities that have arisen from the creation of Lake Nasser, the lake that formed behind the Aswan High Dam in southern Egypt. These stakeholders include diverse groups such as mining companies, migratory fishermen, commercial farmers, governorate planners, politicians, university researchers and last, but certainly by no means least, the resident Bedouin of the region. In fact, it is this last group that arguably has the biggest stake of all, because it is they who live and work here, and use and manage the resources, and it is this group on which this book really focuses. Unlike the other stakeholders, the Bedouin do not have any realistic alternatives, unless permanent migration out of the region to the towns and cities of the Nile Valley itself is considered to be an option. Indeed, for many of their predecessors, especially in the first part of the twentieth century, this was the only option. But it need not be for the present generation of Bedouin in the southeastern part of the Eastern, or Nubian, Desert. This book focuses largely on their story, a story of how the Bedouin of the area have adapted to, and managed, the changing resource base brought about with the construction of the High Dam at Aswan.

The story has two key themes that underlie it: environmental change and sustainability. These are two concepts that can be uneasy bedfellows, and there are frequently tensions between them. Currently, we live in an era of heightened awareness of environmental change at the global level, seen especially in the contested debates on global climate change, or what is popularly known as global warming. Such environmental change is typically characterized as being the antithesis of sustainability; indeed, it is common to see such global climate change as wholly unsustainable. If we transfer these ideas to the local scale, we again come up against tensions between sustainability and environmental change. However, this does not

1

always have to be so, and this book will explore how this need not necessarily be the case.

Part of the difficulty we face emanates from the fact that the term 'sustainability,' or as is more commonly used, 'sustainable development,' is contested. At one level, sustainable development seems straightforward. The Brundtland Commission in 1987 defined sustainable development as "development that meets the needs of the present without compromising the ability of future generations to meet their needs." At face value this seems reasonable enough, but closer inspection starts to generate problems for us. What do we mean by 'needs,' for example? Are the 'needs' of a farmer in sub-Saharan Africa the same as the 'needs' of a New York urban worker? Are the 'needs' of an Egyptian Bedouin the same as an Egyptian fellah in the Nile Valley? Needs are clearly embedded in their cultural and economic settings, and these differ from one place to another. Consequently sustainable development, and what it means, is also going to alter. Geography, it seems, does matter.

There are also issues around what resources are to be sustained. What may seem to be a vital resource in one area may not be so in another. What may be a vital resource at one particular point in time may not be so for the next generation thirty years later. We can take this further by examining different conceptualizations of the environment and its resources. In southern Egypt, can we be sure that urban-based economic planners, for instance, see the resources, and their potentials, in the same ways as Bedouin? And, if not, what implication does this have for how we might define and enact sustainable development on the ground? Sustainability issues dominated both the Rio de Janeiro and Johannesburg earth summits in 1992 and 2002 respectively, but the big difference over the decade was that by the time the Johannesburg summit took place, equity in terms of access to resources had risen up the agenda, to the extent that some now wonder whether sustainable development can ever be achieved without 'environmental justice.' These are some of the questions and issues around sustainable development and sustainability that will be woven into the discussion that follows.

This, though, brings us to the other issue, that of environmental change. In southern Egypt, far-reaching environmental change has been brought about by the construction of the Aswan High Dam during the 1960s. However, this dam is by no means unique. Massive dam projects have been common throughout the twentieth century, across North America, the former Soviet Union, Africa, and Asia. In Africa alone there have been significant dam projects: in Ghana with the Akosombo, in Mozambique, with the Cabora Bassa, in Zambia and Zimbabwe, with the Kariba and, of course, within the Nile Valley drainage basin, with the Gabal Aulia, Sennar, Roseires, and Khashm al-Girba dams in Sudan, and the 1902 Aswan Dam and barrages further downstream in Egypt.

However, it would be fair to say that many, if not all, these projects have been controversial in one way or another. On the positive side, these dams have, to greater or lesser degree, provided major flood controls that have subsequently saved many lives, either directly, by preventing loss of life from catastrophic floods, or indirectly, by saving crops, property, and other economic infrastructure from flood destruction. Some of the larger dam projects have also provided strategic water reserves in their reservoirs that have become central to many countries' development strategies. It can be argued that the Aswan High Dam did much to save Egypt from food deficits during the mid-1980s when Ethiopia, one of the major source areas of the Nile, experienced successive years of failed rains and drought.

On the other hand, there have been costs, not least the financial expense and strain of building such schemes, but there is more. Displaced farmers from lands upstream of such dams have had to be resettled with economic, social, and even cultural implications. Environmental concerns have developed as well. Sediment that would normally have been washed down river channels, and helped to replenish land fertility in the valleys in a natural and rhythmic fashion, is now deposited upstream of the dam wall. Land fertility can only be maintained through increased use of chemical fertilizers with all the implications for soil structure and unused chemicals being washed out into the channels. Despite these concerns, large dam projects continue, and we can see examples of these currently in China with the Three Gorges Project on the Yangtze River and in India on the Narmada River.

In Egypt, the rock-filled Aswan High Dam, which impounds what is called Lake Nasser in Egypt, but Lake Nubia south of the border in Sudan, was constructed between 1959 and 1969. The main purpose of building the High Dam was to create a reservoir sufficiently large that excess water from the Nile's greater flood years could be held in storage and released as and when required for water consumption and for hydro-power generation. The dam was built primarily to conserve water and regulate the river flow; all other uses, except that of power generation, were of secondary importance. The dam itself is 3,600 meters in length, 111 meters in height above the riverbed, and 980 meters in width at the base, tapering to only 40 meters in width at its highest elevation. Its construction regulates the flow of the Nile in Egypt and now provides sufficient storage to supply Egypt's allocated annual share of 55.5 cubic kilometers of water with a high degree of reliability. Before the High Dam was built, about 50 percent of the Nile's water drained into the Mediterranean Sea; after the dam was completed, only an insignificant amount of Nile water (about 0.4 cubic kilometers annually, or only about 1 percent) reaches the Mediterranean except in the extreme high flood years. The rest of the water is used for agricultural, industrial, and domestic needs, as well as being lost by evaporation in natural ecosystems and, increasingly, also from newly reclaimed desert areas.[1]

Both before and since the dam's construction there has been an extensive debate on its impact on the country's environment and on its socioeconomic effects on Egypt, its people, and sectors such as agriculture, industry, tourism, transport, and culture. The new Nile hydrological regime certainly enhanced agricultural production, increased the area of irrigation, changed crop patterns, and altered water schedules. Additional water in Egypt transformed 750,000 feddans of land from basin irrigation, with one annual crop only, to perennial irrigation with multiple crops per year, and enabled the reclamation of one million feddans for farmland.[2] Navigation on the River Nile also improved. In addition, the electricity generated by the dam increased the supply of energy to support the growing industrial sector. This included the manufacture of fertilizer to meet the increasing demand areas of cultivation and to substitute for the loss of the silt formerly deposited by the Nile flood.[3]

The area most affected by the dam's construction was Nubia in the south of Egypt and the north of Sudan, which was submerged by the huge, newly formed reservoir that extended for some 500 kilometers south of the High Dam itself. With the creation of the reservoir, water covered the fertile soils of the Nile Valley in Nubia, where the main cultivated crop had been the date palm; in Sudanese Nubia alone, about 900,000 date palms were grown,[4] and more than 44 villages were lost to permanent inundation by the lake. Approximately 100,000 Nubians were evacuated and resettled in newly reclaimed lands, about half in Egypt and half in Sudan.

Many archaeological sites in Nubia would also have become immersed by the new lake, including the Isis Temple on Philae Island at the First Cataract, about ten kilometers south of Aswan, and the rock-cut temples of Abu Simbel about 280 kilometers south of Aswan. These two monuments were salvaged and successfully moved onto nearby higher ground. Some other free-standing pharaonic temples (Kalabsha, Qirtassi, al-Muharraqa, al-Sebu, and al-Dakka) were also physically transferred to new sites.

The long-term effects of changing the water regime on the transport of silt and sediments that historically built up the fertile Nile Valley soil have been a major concern. These sediments are now trapped in the reservoir, a situation that has led to a decline in the fertility of the Nile Valley soil and of the Mediterranean coastal waters. The latter has caused a severe reduction in the numbers of commercially valuable fish, particularly sardines and crustaceans. Severe erosion along the Egyptian Mediterranean coast has been another harmful outcome of the reduced sediment supply.[5]

It is in this context, therefore, that Bedouins by the Lake examines long-term effects of and adaptation to this major environmental change. The book is not based on a snapshot view of change, perhaps taken a couple of years after the building of the dam. Rather, it draws on twenty years of continuous research on changes in the natural environment and resource base brought about by the impoundment of water behind the Aswan High Dam, as well as on how the Bedouin communities of the area have adapted to and managed these changes. All too often, such people are portrayed as victims of such changes; and regrettably all too often, this is the case. However, to accept this uncritically as the situation in all cases not only misses the complexities and subtleties of these changes and how people adapt to them, frequently to their advantage, incidentally, but also does such people a disservice in that it denies them initiative and a sense of agency. This is not to over-romanticize the Bedouin of southern Egypt; as we shall see, life can be very hard.

The results of these twenty years of research provide us with a very unusual, perhaps even unique, opportunity to examine in detail the natural, economic and sociocultural changes brought about by a major dam scheme. Significantly, in addition to the Bedouin, the researchers themselves have become important players in this story, particularly because they have linked conservation and development in ways that might go some length to encourage sustainable development. Together with traditional scientific studies on the biotic and abiotic components of the ecosystem, the research also emphasizes the role and importance of indigenous Bedouin as essential elements of such systems. The years of multidisciplinary research have contributed to the understanding of the close interrelationships between natural cycles and the socioeconomic trends and characteristics of the area. In particular, the Unit for Environmental Studies and Development (UESD) of the University of the South Valley in Aswan, in cooperation with national and international institutions, has actively promoted interdisciplinary education, training, and research, and field projects in the Wadi Allaqi area by combining social, human, and environmental sciences.[6] The educational approach takes into account the experience of the indigenous culture of the local Bedouin communities living in the Wadi Allaqi area and incorporates it into the process of sustainable development. Significantly, attitudes toward gender have changed, particularly toward the role of women in field research, and this has given female researchers confidence in their ability to work in the harsh desert environment on an equal footing with their male colleagues. In the 1980s there were no Egyptian women carrying out fieldwork in the desert; by the turn of the millennium, half of the UESD team was female. It is not only the Bedouin communities of the area that have experienced, and adapted to, significant changes.

What follows in the book is an account of more than two decades of research results from the Wadi Allaqi area in southern Egypt. Chapter 1 introduces the people of the wadi and the concept and reality of the Wadi Allaqi Biosphere Reserve. Chapter 2 explains the key environmental changes that have come about in Wadi Allaqi as a result of the construction of the High Dam. Clearly, the most significant impact is the presence of a large body of water in this otherwise hyperarid environment, but this has had further implications in terms of the formation of new soils through new patterns of sediment accumulation and loss. The flora and fauna of the area have also responded to this change. Chapter 3 examines the effect of these environmental changes on the resource base for those people living in the desert. This chapter considers the resources traditionally available to Bedouin, and the new opportunities emerging as a result of the High Dam Lake. Chapter 4 considers the impact of these changes on Bedouin livelihoods. After a discussion of the traditional basis of Bedouin livelihoods in the Eastern Desert, the chapter illustrates how the changed resource opportunities have brought new groups of people to the desert (including commercial farmers, fishermen, and miners). These have introduced further changes to Bedouin life, such as improved transport links to Aswan and the emergence of wage labor. Bedouin hold detailed knowledge of the desert environment, and it is this that allows them to manage their environment and survive within it. New resource opportunities are incorporated into indigenous knowledge systems, presenting us with an excellent opportunity to see how these systems deal with new ideas as well as those that are no longer relevant. Importantly, as Chapter 5 demonstrates, these changes are not uniform. Bedouin have been able to take advantage of new opportunities in various ways depending on their relative wealth. Furthermore, these changes have been different for men and women and, as a consequence, their knowledge, and use, of the desert environment has also differed. Finally, Chapter 6 looks to what the future holds for the Wadi Allaqi environment. It will examine the tensions between the desire to develop the resource opportunities in the area with the preservation of this environment, and the daily existence of the Bedouin living there.

1
Wadi Allaqi: The People and the Pressure for Conservation

The Location of Wadi Allaqi

The geographical focus of this book is Wadi Allaqi in southeastern Egypt, located about 180 kilometers south of Aswan (Figure 1.1). Wadi Allaqi itself is the largest wadi of the southern part of the Eastern Desert of Egypt, originating in the Red Sea Hills some 200 kilometers to the southeast of where it meets the Nile valley. In its upper reaches in the hills, the annual rainfall averages less than 50 mm, while in the downstream parts of the wadi, rainfall events are extremely rare and the area is genuinely hyperarid. Even in the upstream parts of the wadi, the figure of 50 millimeters is frequently not reached, and since the mid-1990s there has been a severe drought that has resulted in virtually no rainfall for the last decade or so.

Before the Aswan High Dam was built, Wadi Allaqi was a dry wadi for all its length, at least on the surface (there has always been subsurface water movement flowing toward the Nile). At its mouth, where Wadi Allaqi met the main River Nile, there was located the village settlement of Allaqi, a meeting point for Nubian farmers from the Nile valley and Bedouin nomads from the Eastern Desert. Allaqi village is now submerged beneath the waters of Lake Nasser. Indeed, the main impact on Wadi Allaqi of the impoundment of water behind the High Dam has been the creation of an arm of Lake Nasser extending southeastwards some 80–100 kilometers into the lower part of Wadi Allaqi. This has essentially created a major new resource opportunity—water—for the Bedouin of the area. What was previously a harsh, hyperacid are has now become one where water is guaranteed from the lake itself. Not only that, but grazing resources have become available around the lakeshore.

9

For Bedouin, these factors have become a great attraction. Previous to this the Bedouin economy had to rely on the vagaries of rainfall, particularly in the Red Sea Hills, to produce ephemeral grazing for their livestock. In hard years this was supplemented by using forage materials from trees, especially acacia, but this was, and still is, a relatively scarce resource. The lake now potentially provided a much more secure source of grazing for Bedouin livestock. Furthermore, opportunities for cultivation in pre-inundation times for Bedouin were very limited. Post-inundation, lake water can be used to irrigate small farms, and even at distances away from the lakeshore, relatively shallow wells can be dug to depths of two to four meters, which draw on subsurface water flows associated with the lake.

Figure 1.1: Location map of Wadi Allaqi

The People of Wadi Allaqi

The main group of people who have taken advantage of these new resource opportunities to live around the lakeshore are Ababda, but some Bishari have chosen to settle here also. The Eastern Desert of Egypt is traditional Ababda territory, although the Ababda heartland is located further north to the east of Edfu. The traditional Bishari heartland, on the other hand, is found much further south across the border in northeast Sudan, and so, in a sense, the Bishari of Wadi Allaqi are seen as 'guests' by some Ababda. However, there are few tensions between the two groups. This may be something to do with numbers, in that less than 10 percent of the population of Wadi Allaqi is of Bishari origin. Numerically, the Bishari represent no threat to Ababda, and are certainly not currently present in such numbers as to threaten the resource base of the area.

The Ababda ethnic group comprises a number of sub-groups, which are, in turn, further divided into clans. In Wadi Allaqi the most important sub-group is that of the al-Ashibab, a sub-group that has experienced a significant and steady out-migration from the desert to the adjacent Nile Valley towns and cities of Qena, Luxor, Edfu, Daraw, and Aswan for much of the twentieth century. However, it would appear that the al-Ashibab sub-group has nonetheless experienced rather less of a decline in its desert population than many of the other Ababda sub-groups.

The al-Ashibab sub-group comprises seventeen different clans, of which members from six are found in Wadi Allaqi, but only three are of any numerical, economic and, to an extent, political significance. These three clans are the Sadinab, Hamidab, and Fashikab. Although the Sadinab are the most numerous, especially in the downstream area around the lakeshore with about a hundred people, political influence in the area tends to rest with the Fashikab and Hamidab. With the death of the clan head of the Fashikab in 1999, however, political power has tended to gravitate more toward the Hamidab clan, and many of the Fashikab have decided that their future lies elsewhere around the lakeshore and have moved away for the moment from Wadi Allaqi.

The Bishari who are found in the area originate, or at least did so in the past, from Gabal Elba in upstream Wadi Allaqi and from the Red Sea Hills of Sudan. Similar to the Ababda, the Bishari also comprise a number of sub-groups and clans, and most of the Wadi Allaqi Bishari come from the al-Mallak clan. The Bishari are relatively recent arrivals in the area and certainly postdate the arrival of the Ababda. To avoid conflicts over access to land resources, agreements between each of the three Ababda clans and the

Bishari have been made, and each of the four groups has access to clearly defined and commonly agreed upon areas around the lakeshore. In the event of disputes taking place, of which there have been remarkably few over the years, these are initially resolved at the community level. If agreement cannot be achieved for any reason, the dispute is then referred up to the sheikh with responsibility for Bedouin in Aswan governorate. This person is agreed upon by the Bedouin of the area as someone who can be trusted and who shows good judgment and compassion. He also has the extra authority of being recognized legally by Aswan governorate.

Tensions in Wadi Allaqi

Although tensions between the different Bedouin groups are nonexistent to relatively minor, the same cannot be said for some of the other groups in the area. Wadi Allaqi has a range of potential resources attractive to a number of outsiders. In the 1970s commercially exploitable deposits of talc, granite, and marble were discovered, and these have attracted quarrying interests to the area. The Marnite company is one of these, its name incidentally being an amalgam of marble and granite. Other quarrying interests have also become active in the area.

The lakeshore has been seen as an attractive option for commercial agriculture. In the early 1980s, and subsequently in the late 1990s, the Aswan High Dam Lake Development Authority, charged with the responsibility for the economic development of the Lake Nasser shore lands, offered a range of incentives to would-be developers to invest in large-scale agricultural activities around the lakeshore. Some of these farms were up to a hundred hectares in size and were organized on a commercial scale with the use of heavy machinery and, crucially, the deployment of chemical fertilizers and pesticides. This has had important ecological implications. The water of Lake Nasser is a clean resource, but chemical fertilizers washed into the lake have led to pollution as well as to the accelerated growth of aquatic vegetation and weeds. Furthermore, pesticide residues have been found on the crops produced, especially watermelons, with all the implications that this has for human and animal health. Indeed, some Bedouin have formally complained that they have lost sheep that have grazed on crop waste polluted with pesticides.

In the 1980s there were plans to develop Wadi Allaqi as an agricultural region. The lakeshore vegetation, which had grown in response to the new water opportunity from Lake Nasser, was to be cut down and the land ploughed up and planted with various crops on a commercial basis. Water could be pumped from Lake Nasser through pipelines up Wadi Allaqi and then allowed to flow back down under gravity. Surplus water not used in the growth process associated with crop cultivation would then flow naturally back into Lake Nasser. This would, of course, reduce evaporation losses experienced from unused water standing in the fields, a problem experienced elsewhere further north in some parts of Egypt. However, that water would be loaded with both fertilizers and pesticides, a clear ecological problem for Lake Nasser. Not only that, but if there were to be commercial fields surrounding the lakeshore, then access for Bedouin to the lake for water would be compromised.

Other itinerant groups in Wadi Allaqi include fishermen, mainly from the Luxor and Qena areas in the Nile Valley. Despite the huge fisheries resource presented by Lake Nasser, the Bedouin have little, if any, interest in fishing, and fish are largely missing from their diet. In-migrant fishermen are attracted to Wadi Allaqi, as an arm of Lake Nasser, because of its relatively easy access to Aswan, and especially because of the rich and relatively shallow waters of the wadi that attract fish in large numbers. However, these people put pressure on the resources and exploit the wood resources around the lakeshore for fuel, leading to degradation problems. Given its proximity to the Sudanese border there is also a significant military presence in Wadi Allaqi and the surrounding area, and pressure is exerted by soldiers on the scarce resources, especially woody plants.

The Wadi Allaqi Biosphere Reserve

All of these factors add up to significant pressures on the ecology of Wadi Allaqi, and, as a consequence, the area was formally designated as a Biosphere Reserve in 1993 within the UNESCO Man and Biosphere Program (MAB) and, as such, it represents an extreme arid ecosystem type within the MAB Network of Biosphere Reserves. In such types, care is taken not only to protect biodiversity itself but also to protect any traditional pastoral systems that might be found there, and any indigenous practices involved in managing this biodiversity. The Wadi Allaqi Biosphere Reserve demonstrates the value of conservation in an extremely arid desert region, as well as reconciling of the conservation of biological diversity with its sustainable use.

Although Wadi Allaqi only became a Biosphere Reserve in 1993, it had already received formal protected area status in 1989.[1] The need for the protection of wildlife and biodiversity in Egypt has generally been well recognized by the Egyptian authorities for some time. This recognition has come about because of the decline in biodiversity caused by lost habitats, unregulated hunting, and the overexploitation of economically important plants. The rich wildlife of the Nile Valley and Delta, with their crocodiles, papyrus swamps, and lotus beds, as recorded in pharaonic inscriptions and papyrus documents, is now a feature largely of the distant past. All the Nile Valley land is now cultivated, in some places up to three times a year, leaving only a small patch of semi-natural Nile Valley vegetation, and its gallery forest of acacias, remaining on the First Cataract Islands at Aswan. Widespread hunting in many desert areas has caused a drastic decrease in animal populations, and has even led to the extinction of many mammals. Well-known examples include the declining populations in most Egyptian desert areas of the once abundant Dorcas Gazelle and the Nubian Ibex, now found only in the most remote and inaccessible areas.[2] Extensive and intensive development, including the reclamation of desert land for cultivation, the excavation and mining of minerals, and the construction of new roads, have presented serious threats to wildlife. In accordance with Egypt's policy of protecting biodiversity,[3] representative habitats rich in biodiversity have been declared conservation areas. At present, about 10 percent of Egyptian land now has protected status.

A numbers of key reasons contributed specifically to the declaration of Wadi Allaqi as a protected area in 1989, including its geographical location between two significant biogeographical corridors—the Red Sea to the east and the River Nile to the west. These corridors link the tropics in the south

Figure 1.2: A Nile Crocodile *(Crocodylus niloticus)* with a Spur-winged Plover on the lake shore. Photograph by Dick Hoek.

with the Palaearctic in the north, accounting for the mosaic of ecological systems characteristic of the warm desert biome. Wadi Allaqi as a system spans four distinct ecological formations: the first is the Elba mountain group at the southern, upstream end, seen by some to be a 'coastal mist oasis' that receives orographic precipitation and provides habitats for a rich biodiversity; the second is the main reach of the wadi, extending over part of the 'rainless desert' of Nubia, where only occasional rain (or runoff flow) supports the biotic components of the ecosystem; the third is the ecotone habitat at the deltaic part of Wadi Allaqi, where periodic inundation by the lake causes dense plant growth, including tamarisk bushes and herbs; and the fourth comprises that part of Lake Nasser that extends into the downstream part of Wadi Allaqi. The water that constitutes Lake Nasser provides diverse habitats that enable several species of water animals that have lost most of their habitats elsewhere in Egypt, to regain their populations. These include the Nile Crocodile, the Nile Monitor Lizard, and the Nile Soft-shelled Turtle. These habitats are associated with biodiversity that is of special conservation interest.

Because of the Wadi Allaqi area's remote location and difficulties of access, its wildlife and flora remained relatively undisturbed until the 1970s. Considering that vegetation occurs in only very limited areas of the narrow wadi channel, the floristic diversity is rich, with some 127 species having been recorded; these belong to thirty-eight families, with most represented

Figure 1.3: A Monitor Lizard *(Varanus niloticus)* on the shore of Lake Nasser. Photograph by Dick Hoek.

by either one or two species. Despite there being no endemic species, a few plants present in this area are very rare in the Egyptian flora, among them the grass *Cymbopogon proximus* (*halfa barr* in Arabic), commonly used in medicine. Large undisturbed populations of *Salvadora persica* shrubs (*arak* or *miswak*) and *Balanites aegyptiaca* trees (*higlig* or *lalub*) still exist in the upstream part of the wadi, but are threatened elsewhere in Egypt. However, in establishing priorities in conservation the number of species is not an infallible measure, or even necessarily a moderately good indicator of biotic diversity.[4] On the contrary, an area so large, but with such a limited resource base, should have a high priority for fear of loss of the invaluable that is still available. Most of the plants growing in Wadi Allaqi are used by the inhabitants and provide shelter, as well as fodder for livestock, fuel, and medicine for the human population. They also provide a source of income when sold in the market. Ephemeral plant life in the channels of Wadi Allaqi and its tributaries is 'accidental,' in that rich plant growth may appear in a year, followed by a series of years with no rainfall and no, or very little, plant growth. This situation limits the species that can survive this water availability regime to those species whose seeds may remain viable for up to ten years or even more. This is an attribute of special significance that is very important for genetic conservation.

Wadi Allaqi formerly had important populations of some nationally rare and endangered fauna species, such as the ostrich *Struthio camelus*. Until the 1960s ostriches were common in the southern part of the Eastern Desert, according to the Bedouin and records of other travelers. However, during intensive studies of the Eastern Desert in the twenty-six-year period from 1980 to 2006, only one group of four ostriches was recorded in the upstream part of Wadi Allaqi, this being in 1991; otherwise there have been no other recorded sightings of ostriches in the area since 1980. Wadi Allaqi still has one of the largest Dorcas Gazelle *(Gazella dorcas dorcas)* populations remaining in Egypt, and a small, but important, population of Barbary sheep *(Ammotragus lervia)*. Dorcas Gazelle are common visitors to the lakeshores, especially during the dry periods when grazing is scarce in the upstream part of the wadi. Sometimes they eat crops growing on Bedouin farms, with a particular preference for fleshy vegetables such as ladies' fingers. Many foxes of both Red Fox *(Vulpes vulpes)* and Rüppell's Sand Fox *(Vulpes rueppelli rueppelli)* species are resident in the tamarisk thickets. Striped Hyena *(Hyaena hyaena)* and jackals *(Canis aureus)* are also present in small numbers in Wadi Allaqi. The Cape Hare *(Lepus capensis)* is very common in the downstream part of the wadi, and likes to feed on the branches of small *Balanites aegyptiaca* trees that are grown on the Wadi Allaqi Project experimental farm. Among rodents, two small gerbils—*Dipodillus campestris* and *D. mackilligini*—were recorded by L. Emelianova, a Russian zoologist working in this area in 2001.

Wadi Allaqi is also situated on one of the main migration highways for birds moving between the warm tropics and the temperate regions. Lake Nasser has become an important resting and watering area for wintering migratory birds, and Wadi Allaqi, especially in its downstream part, provides a habitat that receives rich populations of birds in both winter and spring, placing the lake's wetlands on the list of Egypt's most important bird areas.[5] In the downstream part, the riverine environment is suitable for several resident bird species, such as the Moorhen, Purple Gallinule and Pied Kingfisher, Lake Nasser having created suitable breeding conditions for them. Unfortunately, Egyptian Geese, whose population has proliferated in this area, causes great damage there to new fields, therefore creating a serious conflict with conservation officials. Flamingo and pelican passage migrants seem to have increased on the lakeshores in Wadi Allaqi in recent years. A few individual flamingo and pelican were seen in the wadi on the shore of Lake Nasser in the 1980s, and their numbers have subsequently increased. Since 2001, large flocks of more than two hundred individuals of pelicans and flamingos have been seen regularly in autumn on the open water in Wadi Allaqi. The wadi is an important resting point for the White Stork on its annual migration, especially returning to its northern breeding

Figure 1.4: A rarely-seen jackal *(Canis aureus)* prowling along the lake shore. Photograph by Dick Hoek.

Figure 1.5: A flock of White Pelicans *(Pelecanus onocrotalus)*—winter visitors to Lake Nasser. Photograph by Dick Hoek.

grounds in the spring, when many thousands are seen resting in the wadi. Storks have been observed feeding on locusts, thus naturally protecting the wadi's vulnerable vegetation.

However, it is not only Wadi Allaqi's biodiversity, but also its cultural significance that highlights the importance of this area for protection measures. The ancient caravan route between Sudan and Egypt through the wadi was a shorter and easier journey than following the Nile itself. The fortresses of Bak (or Kuban) at the mouth of the wadi were built as far back as the Old Kingdom, when Wadi Allaqi was a key contact zone for Egyptians and the tribes of the Nubian Desert and played a very important role in the life of this part of the desert. Inscriptions dating back to the Fifth Dynasty confirm that this region was not only used as a caravan route, but also that stone for sarcophagi was quarried here. Indeed, the 1995 Institut Français d'Archéologie Orientale (IFAO) archaeological survey of Wadi Allaqi and its tributaries clearly demonstrated the rich archaeological potential of the region, which has suffered comparatively little interference or modern disturbance. Signs of Neolithic occupation (tombs, settlements, and rock carvings) have been found in different parts of the Allaqi basin, and notably in Wadi Gabgaba.[6] These show the presence of populations that were culturally distinct from the Nile Valley tradition and prove that the area has been occupied since ancient times. Related to this is Wadi Allaqi's mineral wealth, especially of metals and ornamental stones. Since ancient times, marble and granite, together with metal ores including gold and copper, have been extracted and either used or sold, both within and outside Egypt. Especially important are the gold mines, remains of which are seen in almost every large wadi, but all have ceased mining. In Wadi Allaqi seven large old gold mines and numerous small ones have been recorded, the largest of which, Umm Qariyat gold mine in Haimur, closed down in the 1930s.

Given this background and the building pressures and tensions arising from the increased human use of the new resource base created by Lake Nasser, it was no surprise when Wadi Allaqi became a UNESCO Biosphere Reserve. The origin of the Biosphere Reserves concept goes back to the Biosphere Conference organized by UNESCO as far back as 1968. This conference highlighted on Biosphere Reserves as alternative ways of conserving the ecosystems, by putting emphasis on the sustainable uses of natural resources in the interests of the people that lived both within and around them. Such a program, initially named Man and the Biosphere (MAB), was officially launched by UNESCO in 1970. One of the key projects (particularly project number eight, "Conservation of natural areas and of the genetic material they contain"),[7] within the MAB program was concerned with the establishment of the world network of Biosphere Reserves, which were to represent the main ecosystems of the world in which genetic resources would be protected, and where research on ecosystems, as well as monitoring and training work, would be carried out for an intergovernmental program called for by the conference. The Biosphere Reserves strategy, which was primarily more scientific than developmental, was substantially revised in 1995, with its adoption by the UNESCO General Conference of the Statutory Framework and the Seville Strategy for Biosphere Reserves. The link between the conservation of biodiversity and the development needs of local communities, a central component of the biosphere reserve approach, was recognized by the Seville Strategy as a key feature in the successful management of most national parks, nature reserves, and other protected areas.[8] Modified approaches to biodiversity conservation include integrated conservation-development projects, which attempt to reconcile conservation needs with the needs of local people and particularly the involvement of the latter in the management of protected areas.

A Biosphere Reserve is designed to deal with one of the most important questions facing the world today: how can we reconcile the conservation of biodiversity and biological resources with their sustainable use? Biosphere Reserves are under national sovereign jurisdiction, yet share their experience and ideas nationally, regionally and internationally within the World Network of Biosphere Reserves (WNBR). There are 507 sites in 102 countries world-wide within the WNBR, which provide context-specific opportunities to combine scientific knowledge and governance modalities; to reduce biodiversity loss; to improve livelihoods; and to enhance social, economic and cultural conditions for environmental sustainability. Biosphere Reserves can also serve as learning and demonstration sites within the framework of the United Nations Decade of Education for Sustainable Development (DESD).[9]

Among many MAB definitions, the most relevant to the situation of Wadi Allaqi is as follows: "Biosphere reserves are sites which seek to reconcile socioeconomic development, and conservation of biodiversity based on local community efforts and sound science. They constitute ideal places to test and demonstrate new approaches to sustainable development at a regional scale providing lessons which can be applied elsewhere."[10] Biosphere Reserves serve to combine three functions: conservation—contributing to the conservation of landscapes, ecosystems, species, and genetic variation; development—fostering economic development that is ecologically and culturally sustainable; and logistic—support for research, monitoring, training, and education related to local, regional, and global conservation and sustainable development issues. It is the synergetic combination of these three functions that characterizes a Biosphere Reserve.

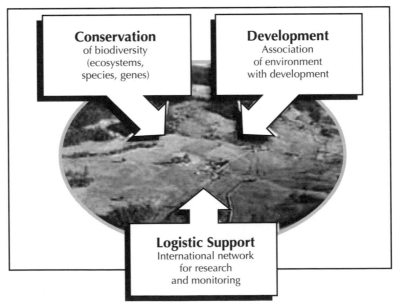

Figure 1.6: The three functions of Biosphere Reserves.

To carry out the different activities involved in these three functions, Biosphere Reserves are organized into three interrelated zones. A core area is legally established and state-owned to ensure long-term protection. There can be more than one core area, but the core should be large enough to meet its conservation objectives. There is minimal human activity. A buffer zone surrounds, or is located next to, the core. This can be an area for experimental research to enhance high-quality production, while conserving natural resources or ecosystem rehabilitation. It can accommodate education, training, tourism, and recreation facilities. The third zone is an outer transition area, whose borders are not necessarily fixed. It is here that the local communities, nature conservation agencies, scientists, cultural groups, private enterprises, and other stakeholders must agree to work together to manage and develop the area's resources on a sustainable basis for the benefit of the people who live and work there. In accordance with these principles, the Wadi Allaqi Biosphere Reserve is organized into the standard three MAB interrelated zones: the core area, the buffer zone and the transition area. There are, however, two core areas and two buffer zones within the reserve, which are located in the downstream and upstream parts of the wadi, separated by a distance of approximately 150 kilometers.

Table 1.1:   Areas of Wadi Allaqi Biosphere Reserve (square kilometers).

	Area (km²)
Total	23,100.00
Core areas	
Quleib core area	350.50
Eiqat core area	235.50
Buffer zones	
Quleib buffer zone	555.75
Eiqat buffer zone	409.25
Transition area	21,549.00
Altitude (meters above sea level)	+165 to +1,500

Note: Size of terrestrial core area(s): 586 km².

The first core area is named Quleib, and comprises the floors of Wadi Quleib, Wadi Umm Arka, and Wadi Heisurba, all of which are downstream tributaries of Wadi Allaqi (Figure 1.7; Table 1.1). This core area comprises a compact, representative section of the true desert wadi ecosystem typical of the southern part of the Eastern Desert. Vegetation is very sparse, and plant communities are dominated by acacia trees (*Acacia tortilis* subsp. *raddiana* and *Acacia ehrenbergiana*) and a few drought-resistant perennials *(Aerva javanica, Senna alexandrina)*, as explained in more detail in Chapters Three and Four. The spatial pattern of the permanent vegetation is largely related to sub-surface water flows and the configuration of the flood channel, and the temporal pattern is linked to rare flood events and the availability of a seed bank from which the ephemeral component of the vegetation cover develops after any rain or flood. A large population of Dorcas Gazelle is present in this area, even during rainless periods.

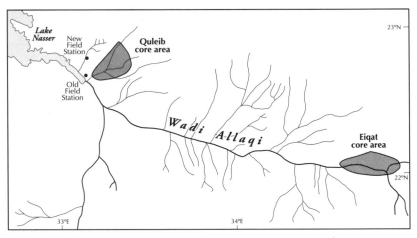

Figure 1.7: Map of Wadi Allaqi Biosphere.

In the period when Lake Nasser's water level was rising between 1998 and 2002, water from the lake entered into the downstream part of Wadi Quleib forming a narrow inlet. Subsequently, an ecotonal system, similar to that formed in the main Wadi Allaqi, was formed in the limited area at the mouth of Wadi Quleib, which was subjected to lake inundation. The zonation pattern of vegetation is characteristic of the ecotonal system, where annual plants (short grasses and *Glinus lotoides*) on the shores are replaced by perennials (tamarisk and *Pulicaria crispa*) with increasing distance from the lake and increasing period of time after inundation.

To avoid lake inundation, part of the Bedouin population moved out of Wadi Allaqi itself and into Wadi Quleib in the late 1990s, together with their livestock, and relocated their settlements in close proximity to, or even inside, the core area. As a result grazing pressure has increased on the surrounding desert system and in the core area. Because of the flooding access to the lakeshores has become increasingly difficult; one of the more readily accessible routes to the lakeshore for water collection from the lake is through the core area in Wadi Quleib. Vehicles damage the existing plant communities, thereby destroying the seed bank and causing soil erosion.

The second core area, Eiqat, lies in the upstream part of the Wadi Allaqi basin and is named after Mount Eiqat, which rises to some 1,400 meters ASL. The wadi floor itself is at an elevation of approximately 500 meters ASL. The Eiqat core area comprises a representative section of the wadi ecosystem typical of the Red Sea Hills in the southern part of the Eastern Desert. In

Figure 1.8: A Bedouin settlement in the downstream part of Wadi Allaqi.
Photograph by Irina Springuel.

general, ecological conditions are determined by the extreme aridity of the area, although it is more humid than downstream because of the prevailing influence of the Red Sea and the more frequent rain and water runoff events from the hills. The Eiqat area is richer in species diversity than the downstream part of Wadi Allaqi. About 100 vascular plant species have been recorded, among which are many rare species of high economic value. Vegetation is represented by open scrubland, dominated by acacia, *Balanites aegyptiaca*, and *Salvadora persica*. The area is rich in mammals, particularly the Dorcas Gazelle, of which a large population now exists. Hyraxes, which live in the more rocky environments, can frequently be seen at sunrise and at sunset. It is possible that the rocky hills, which are difficult to reach and still largely unexplored, provide a refuge habitat for cheetah; local Bedouin claim to have seen cheetah here as recently as the 1960s.[11] Occasionally, ostriches have been seen in this area, the only remnant species of a more extensive population that used to be common throughout the Egyptian deserts. Despite the presence of only a few species of bird, the desert open woodland in the upstream part of Wadi Allaqi provides a refuge for two very rare resident breeding birds. One is the Fulvous Babbler, formerly resident in the Nile Valley from south of Aswan to the Sudanese border, and the other is the African Collared Dove, distinguished by its beautiful pink head and its peculiar call.[12]

The main function of the core area is the protection of biodiversity, with minimal or no human activity. However, with the modified approaches toward the role of people in the Biosphere Reserve and participation of local people in its activities, the current philosophy for the core area is being reconsidered. If the activities in the surrounding buffer and transition zones are directed toward sustainable development, where the concept of conservation is integrated, the importance of the core area (especially its protection, by being fenced off) is questionable. Furthermore, with regard to rangeland Biosphere Reserves such as Wadi Allaqi, where the lifestyle of the Bedouin is consistent with biodiversity conservation, and where they indeed see themselves as part of the ecosystem, grazing cannot, and should not, be prevented in the core area, since such land management would negatively affect Bedouin livelihoods by depriving them of traditional grazing rights.[13] It would also result in a decreased, or even lost, sense of responsibility for protecting the environment in which they live, because of external decision-makers taking over responsibility for the environment. The prevention of grazing in core areas can only destroy trust and create problems between Biosphere Reserve managers and the local population.

Grazing can be considered to be an inherent, even natural, land-use process in dry land areas, a practice that has taken place for millennia. Selective grazing and browsing control the dynamics of plant communities. Taking into consideration that pasture is available only after very infrequent rainfall events, which happen only once every few years in the Wadi Allaqi Biosphere Reserve, there is no danger of overuse of resources. Particularly for the Eiqat core area, grazing cannot become a problem because Bedouin have their own strategy for the conservation and management of areas without having a destructive impact on the environment. In the Quleib core area, however, some damage has occurred as a result of an increasing population of Bedouin in close proximity to the shore of the lake and their tendency to become semi-settled in this area.

Surrounding the two cores are two buffer zones. In Wadi Quleib, the boundaries of the buffer zone are determined by the watershed of the wadi basin. This is the desert plateau largely devoid of plant cover, but with narrow wadis where a few annuals grow after rain. The ecotone, comprising that part of Wadi Allaqi periodically subjected to inundation by Lake Nasser, is also included. However, the ecotone is not stable and the rich plant biomass is temporal, depending on seasonal and annual fluctuations of the lake water. Research work and a careful monitoring of the ecotonal system have been carried out in this buffer zone. An experimental research station, along with some demonstration plots where indigenous plants have been grown, are located in this area. These experimental sites were flooded and partly destroyed during the inundation peak from 1998 to 2002.

In the Eiqat area, the buffer zone boundaries are determined by the Sudanese border to the south and an area around the core to a distance of about ten kilometers from the core boundaries. Since this is a mountainous zone, it is dissected by numerous deep wadis. Biodiversity is rich and very similar to that of the core area. There is a temporary settlement of Bishari Bedouin just on the Sudanese border comprising about ten huts, the temporary home of twenty-five to thirty people. The site of this settlement, to which the Bedouin periodically return, was determined by the location of a well and the available water contained in it. People come to live here only in dry periods, while for the rest of time they take their livestock to ephemeral pastures. The activities in this buffer zone are complementary to the traditional economy, in which grazing and the collection of medicinal plants are the main components. Charcoal, also, is occasionally produced from dead acacia trees, mainly for local use; commercial production is not profitable because of transport difficulties.

Finally, the transition zone comprises the rest of the entire Wadi Allaqi basin, which is not made up of core and buffer zones, from Lake Nasser to the Sudanese border. Much of this area is characterized by a desert plateau dissected by wadis. The area is large and supports traditional rangeland activities. The wadi vegetation is used by Bedouin for intermittent grazing, the collection of medicinal plants, and the production of charcoal when conditions are suitable. The desert plateau is rich in minerals of which the most valuable are extracted. Significantly, any commercial development of Wadi Allaqi is legally only allowed in the transition zone.

A program for habitat restoration, particularly the development of planted vegetation, as an alternative to the exploitation of the stock of wild plants is currently being explored in this zone. The transitional area is the development zone where researchers and managers associated with the Wadi Allaqi Biosphere Reserve can apply the results of sustainable development trials and experiments. This is also the area where local Bedouin communities, environmental agencies, scientists, private companies, and other stakeholders are required to agree to manage and develop the area's resources on a sustainable basis for the benefit of the people that live and work there.

Much of the formal management strategy in the transition zone is concerned with the promotion of pastoral land use based on traditional, or indigenous, knowledge; the development of 'farms' of economically important plants as an alternative to the exploitation of wild stocks; and the support of small-scale cultivation as a realistic and sustainable alternative to large-scale farms on the lakeshores. The excavation of mineral resources has been directed to areas where mines cannot damage the environment or the landscape of the wadi. Ecotourism is another activity that is being explored as a realistic economic possibility in this zone.

Within the Biosphere Reserve, the Wadi Allaqi system still provides a livelihood habitat for its nomadic inhabitants, as discussed in Chapter 4. The pattern elsewhere in Egypt has been one of decline as a result of government policies of land ownership and sedentarization of nomads.[14] Nonetheless, it is still a very hard existence for the Bedouin of the area. When, in November 2001, the Governor of Aswan visited the Allaqi area and held a joint meeting with local Bedouin communities, Wadi Allaqi Biosphere Reserve administrators, and Egyptian Environmental Affairs Agency (EEAA) consultants, the Bedouin representatives asked that their living standards and facilities be improved. The governor decided to build a village for local inhabitants to provide them with houses and social services, particularly some education and health facilities. The location for this village was selected to be on the boundary of the conservation area, near the foot of Wadi Umm Ashira. Construction was completed in 2003. Despite the Bedouin choosing not to live in the new houses, still preferring their traditional tents, sedentarization seems to be taking place and appears to be proceeding quite quickly and smoothly. Consequently, new economic alternatives for the local people are essential to reduce over-exploitation of the areas surrounding the settlements in the conservation area.

The available and reliable lake water and rich biomass in the downstream part of Wadi Allaqi has attracted numerous developers who look for short-term economic benefits, and are prepared to risk the destruction of the fragile ecotonal system and the potential pollution of the lake itself to gain these benefits. In this respect, the ecotone protects the shoreline of the lake, forming a buffer zone between desert and lake. The dense tamarisk growth prevents invasions of aquatic weeds, such as reeds, on the lakeshores, and protects the lake from migratory sand dunes, especially on its western side. However, this zone could become a main source of lake water pollution through the mismanagement of its natural resources by destroying the plant cover or by inappropriate development such as large-scale agriculture and industry. This ecotonal system is unique not only in Egypt but also in the world's arid lands, and hence special measures for the sustainable use of its resources need to be supported by research and monitoring on the one hand, and by deploying the traditional or indigenous knowledge systems of the local population on the other.

Given the highly valuable conservation components of the area, particularly the biodiversity of an extreme arid biome, and the culture of its pastoral people and their traditional forms of land use that demonstrate a balanced relationship between people and nature, it is clear that Wadi Allaqi meets, even exceeds, Egyptian national priorities and expectations for a protected conservation area. Moreover, it can be argued that these conservation parameters more than meet the criteria of Biosphere Reserve implemented within the UNESCO Man and Biosphere Program (MAB), thus justifying the creation of Wadi Allaqi as a Biosphere Reserve in 1993.

Wadi Allaqi Biosphere Reserve Functions

For the Wadi Allaqi Biosphere Reserve specifically, there are three fundamental objectives to its conservation management program. These are to protect genetic resources; to sustain actual and potential biological resources but in the context of conserving traditional land use systems; and to analyze the effects of continuing environmental and human impacts associated with Lake Nasser. To achieve these goals, the Biosphere Reserve is intended to fulfill three complementary functions: conservation, development, and logistics. The plant and animal components of biodiversity are simultaneously to be saved and studied, leading to their sustainable use and the education of students and the public, thus increasing awareness of the importance of biodiversity in order to gain popular support for conservation. In the functioning (conservation-development-logistics) of any Biosphere Reserve, and Wadi Allaqi is no exception, different actors are involved, all playing their own role in its management.[15] Conflicts of interest in the

Biosphere Reserves are to be debated by all the stakeholders concerned. These stakeholders include government authorities responsible for development generally; private companies working in the area; national and international conservation organizations; government leaders; researchers and educators; the military present in this area; and last, but by no means least, local Bedouin communities.[16]

The role of the protectorate staff is to manage the reserve to maintain its healthy ecosystems. Both biotic and abiotic components of the ecosystem (plants, animals, habitats) are protected, together with the traditional ways of life of local people, along with the scenery and beauty of the area's landscape. Because green plants comprise the essential component of the ecosystem, supporting animal and human consumers, special attention is paid to the conservation of economically important plants. Species that are rare, but intensively used, have the highest conservation priority. Examples of such species include *Balanites aegyptiaca,* all acacia species found in the area, *Ziziphus spina-christi* (*nabaq* in Arabic), and plants of significance for medicinal and grazing purposes, which are described in Chapter 3. Accordingly, one of the main strategies for the protection of biodiversity in the Wadi Allaqi Biosphere Reserve is based on the sustainable use, and the reestablishment where needed, of economically important plants that have become rare and/or are potentially threatened by overexploitation.

Wadi Allaqi is rich in minerals, and the exploitation of these provides considerable economic benefits for Egypt as well as employment opportunities in quarrying and mining. The mining of minerals can be very destructive, however, with both direct and indirect impacts on the surrounding areas. The quarrying of rock and the deposition of waste material can cause land degradation by destroying habitats, as well as by damaging the aesthetic beauty of the landscape. The cutting of trees and shrubs for fuel has a clear and direct impact on the vegetation; only a very sparse growth of woody perennials is now usually observed in close proximity to the mines and quarries of the area.

Digging wells and using underground water for domestic purposes, as well as for processing the ores, threatens the available water resources, which will, in turn, negatively affect plant growth. The construction of tracks and roads in connection with the exploitation of the mineral resources causes additional adverse effects on the surrounding environment, especially through seed-bank destruction. Environmentally safe technology for the recovery of metals from ores is not widely used, but should be taken into consideration in light of government plans for the reopening of old gold mines in the southern Eastern Desert, and particularly in the Wadi Allaqi Biosphere Reserve.[17] Special rules are applied for operating mines and quarries, rules that are monitored carefully by the staff of the Biosphere Reserve, who regularly hold meetings with the miners.

Training programs for students and staff working in the Wadi Allaqi Biosphere Reserve include methods of fieldwork, monitoring, communicating with local communities, and establishing a database. The training of rangers and researchers of Biosphere Reserve staff, who are university graduates but do not always have adequate environmental knowledge, has improved their ability to implement the reserve's management program and conduct their own research work. Postgraduate students' environmental research projects are constantly being incorporated into the program of education for sustainable development. The central focus of this program is on the sustainable management of natural resources for development in marginal and fragile environments. To be accepted for implementation in the Wadi Allaqi Biosphere Reserve, projects should fulfill the main goal of sustainable development by identifying the relevant problems, conducting research, and providing recommendations on how to solve them. Crucial to this program is the inclusion in the research, not only of indigenous

Figure 1.9: A mixed class in the literacy program given in Wadi Allaqi by women members of the UESD team. Photograph by Irina Springuel.

knowledge systems, but also the direct incorporation of inhabitants' attitudes to needs and priorities of development and equitable growth in the area. Sixteen research projects combining science and social science were implemented in Wadi Allaqi between 1987 and 2005, and were conducted by multidisciplinary teams of Egyptian and overseas researchers. These included the following projects: Environmental Valuation and Management of Plants; An Assessment of the Medicinal Value of Plants; The Cultivation of Indigenous Plants; and The Role of Bedouin Women in the Household. The results of this research have been transferred to the protectorate staff, developers and the decision-makers who have interests in the area. Through these various collaborative initiatives at national, bilateral and international levels, research and training activities cover a wide range of issues related to arid zone ecology and resource use.

The key actors in the Wadi Allaqi Biosphere Reserve, however, and central to the whole UNESCO Biosphere concept, are the local inhabitants who use the natural resources for maintaining their livelihoods by combining traditional knowledge, conventional wisdom, and modern science, where each of these has relevance to their everyday livelihood activities. Each activity has its own management strategy to ensure that natural resources are conserved and used in a sustainable way. Even such a potentially destructive activity as

Figure 1.10: Gar al-Nabi, the late Sheikh al-Ababda in Wadi Allaqi, talking about Bedouin life. Photograph by Gordon Dickinson.

charcoal making is managed carefully, as discussed in Chapter 3, to ensure that the vegetation stock is not degraded.

Existing opportunities for the sustainable development of rangeland are quite limited unless entirely new products can be created and given a market identity.[18] In dry rangeland areas three kinds of development are actively promoted to improve the incomes, and hence living standards, of local populations. The first, traditional to the Wadi Allaqi region, comprises livestock herding and its by-products, while the second comprises the cultivation of novel crops such as medicinal plants or fodder. The third kind of development is very new and untried, and involves the promotion of eco-tourism, handicrafts, and 'cottage' industries, producing oils, for example, from indigenous desert plants, notably from *Balanites aegyptiaca*.

To achieve the main goal of the MAB program of the sustainable use of natural resources in the interests of the people who live in a Biosphere Reserve, the local people have to become involved in activities. There are a few principal ways in which Bedouin living in Wadi Allaqi can partici-pate. Information gathering is an important activity where local people are involved through their indigenous knowledge of both biotic and abiotic components of the ecosystem in which they live, as well as of traditional land use and management. In the study of the floristic diversity of the Wadi

Figure 1.11: Bedouin elders and younger men visit the UESD field station in Wadi Allaqi for a regular meeting with the research team. Photograph by John Briggs.

Allaqi, as well as other desert areas, researchers often refer to the knowledge of local inhabitants, who act as guides in the remote desert areas. The local names of the plants, their morphological characteristics, auto-ecology, habitat characteristics, and main uses are well known to the local people, a knowledge that they are usually prepared to share with researchers. A scheme for the assessment of the economic and conservation value of plants found in the Allaqi region has been developed on the basis of such indigenous knowledges, with the help and participation of the indigenous Bedouin.

Another example where local people have participated in research has been in assessing the economic value of acacia trees in the Wadi Allaqi Biosphere Reserve.[19] By comparing uses of acacia for charcoal production (dead trees), and as fodder for livestock (living trees), it was found that the latter brings a higher household income than charcoal through the sale of animals to which it has been fed, even though it is normally used only as drought-reserve fodder and therefore only at those times when other animal feed is very scarce. This explains why resource conservation, such as that of the acacia trees, is a concept inherent in the Bedouin livelihood and value system.

Local people have also been consulted about what plants they would like to have cultivated in the buffer and transition zones, and have suggested fodder crops followed by vegetable and medicinal plants. They are also involved in decision-making; deciding when and where to take livestock for grazing in the Biosphere Reserve is undertaken strictly by local people. There is no influence from the reserve managers, as long as overgrazing does not occur. Bedouin are also responsible for the location of their farms, provided that they are not located in the core or buffer zones. On the other hand, decisions concerning what indigenous plants should be cultivated on the experimental farm are made by researchers in consultation with Bedouin, and suitable locations are identified by Biosphere Reserve staff and researchers. Some local people work in the experimental plots and others help rangers in patrolling the area, acting as guides on tours of remote areas of the reserve.

2
Wadi Allaqi: Changing Environments

The Aswan High Dam Lake

As we have seen, the Aswan High Dam Lake is the name given to the entire reservoir behind the Aswan High Dam, that part of the reservoir within Egypt being named Lake Nasser, while the Sudanese part is called Lake Nubia. This extensive reservoir extends upstream from the Aswan High Dam between the latitudes 23° 58′ and 20° 27′ N and longitudes 30° 35′ and 33° 15′ E (Figure 2.1). The lake is 495.8 kilometers long—at a lake level of 180 meters above sea level (ASL)—and narrow, with its variable shoreline contour determined by the immediate topography. The ratio of shoreline to the length of the entire reservoir is 18:1.[1] The entire reservoir has a gross capacity of 157 cubic kilometers. The 'dead storage' is about 30 cubic kilometers (some 26 percent of the capacity) at 147 meters ASL, which is also the minimum level for operating the hydro-electrical power station. The characteristic feature of this reservoir is its location in an extremely arid climate with very low precipitation of about 4 mm per annum, high evaporation of about 3,000 mm per annum, and summer temperatures that can exceed 40ºC. In such conditions, it is not surprising that some 10–11 percent of reservoir water is lost by evaporation annually.[2]

The lake is a very dynamic system, and its parameters are influenced by the amount of water stored in it. Variations in the size and shape of the lake are affected by the volume of the annual inflow, which depends on both the upstream Nile flood and the outflow controlled in accordance with the prevailing policy for the release of water from the dam. Almost 85 percent of the Nile flow to the lake is water from the Blue Nile and Atbara rivers. Both originate on the Ethiopian plateau and are characterized by their torrential nature, carrying violent floods typically the months from August until October. Accordingly, there can be significant fluctuations in the water flows in different years and at different seasons.

Figure 2.2 shows that the annual fluctuation of the lake water since the formation of the lake in 1963 has expanded and contracted, not only

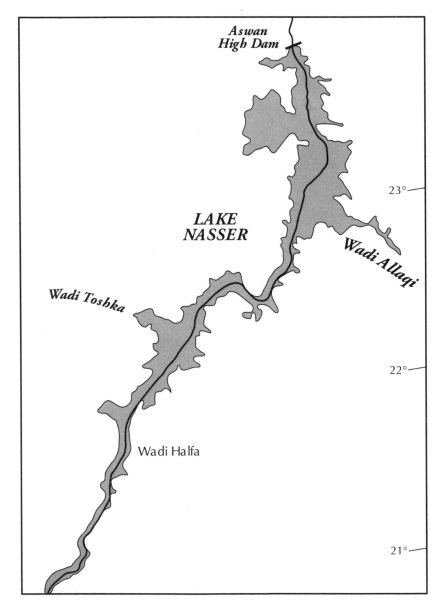

Figure 2.1: The High Dam Lake.

seasonally with the rise and fall of the flood, but varying enormously in extent in different periods. A fluctuation in the water level of as much as thirty-one meters has been recorded, which has led to considerable variations in the size of the lake's surface area.

Table 2.1: Morphological data for the Aswan High Dam Lake.

Water level	160 m ASL	180 m ASL
Length (kilometers)	430.0	495.8
Surface area (square kilometers)	3057 3074*	6216 6220*
Volume (cubic kilometers)	65.9	156.9
Shoreline (kilometers)	6027 5416*	9250 7875*
Mean width (kilometers)	7.1	12.5
Mean depth (m)	21.6	25.2
Maximum depth (m)	110	130

Sources: Entz 1976; * Raheja, 1973.

The reservoir began to fill with water in 1963, and by 1978 it was almost at full capacity. The volume began to decrease again in 1981, and the lowest level so far recorded was 150.60 meters ASL in July 1988, which corresponds to a total storage of 38.4 cubic kilometers—only about eight cubic kilometers above the dead storage level. This was followed by a sudden rise of the water level, which subsequently reached its second highest level (since 1978) in 1999, of above 181 meters ASL, almost reaching the maximum projected level of 182 meters ASL. The water level in the lake remained high throughout 2000, even during the summer months. Since 2001, the water level has once again been continuously but steadily dropping.

When the water level in the lake rose above 180 meters ASL between 1996 and 2000, to avoid any hazards the water spillway began operating and water was allowed to enter the Toshka outlet, which leads water to the chain of Toshka depressions located in the Western Desert. Because of the morphology of these depressions, this water formed huge pools within the Western Desert and silt brought by the water covered the depression beds. Parts of a previously almost waterless desert were transformed into a temporary aquatic, wetland system. When the level of the water in the lake started

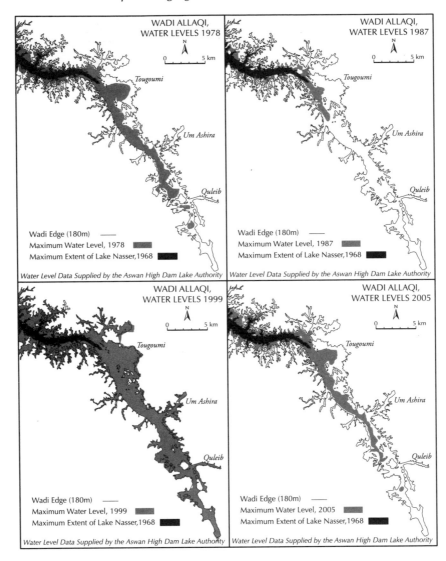

Figure 2.2: High Dam Lake water levels in 1978, 1987, 1999 and 2005.

to decline again after 2001, the spill ceased and these enormous pools began to shrink, driven not only by the lack of further water inflows but also by water evaporation and seepage.

Before the construction of the dam it was estimated that the mean annual deposition of silt in the reservoir would be of the order of 110 million tons, distributed throughout the length of the Aswan High Dam Lake.[3] However, the sediment load is being deposited mainly in the vicinity of what was the Second Cataract in Sudan and is considerably less to the northern end of the lake in Egypt. By 2000, it is estimated that about 134 million tons of silt were being deposited annually within Sudan, most of it south of Wadi Halfa where a new delta is forming.[4]

With the filling of the reservoir, concern was expressed about the possible seismic effect associated with the weight of the water in the reservoir on the underlying rock as well as on the dam itself. In Egypt, an earthquake of 5.3 on the Richter scale was recorded in the locality of Lake Nasser in November 1981, provoking intensive studies of the seismicity of the region. These studies revealed that the earthquake was caused by tectonic activity and that "the dam would remain stable in the face of the most severe earthquake that the area might be expected to suffer."[5]

The quality of water in the reservoir differs markedly from that of running river water. The reservoir is mainly lacustrine, except in its central and southern parts, which have remained riverine. Its water is poorly mixed and exhibits a distinct stratification pattern for most of the year, the stratification of shallow waters in the inlets *(khor)* being different from that of the main channel. The lake has alkaline water (pH about eight) and a high concentration of nutrients, attributable to the high evaporation rate. There are notable differences in water characteristics between the southern and northern sections of the lake; a decrease in phosphates, chlorides, sodium, potassium, silicon, and turbidity, and an increase in sulfur, have been observed in the northern part of the lake.[6] Although most limnological attributes indicate good water quality, there are some elements that could lead to ecological problems.[7] From 1987 to 1992, water bloom, caused by the overgrowth of *Cyanophyceae* species, occasionally occurred in the southern part of the lake. At present, water bloom recurs throughout the year but in different parts of the lake, making geographical predictions of where such blooms may occur very difficult.

Lake Nasser is an important source of national fish production for Egypt. However, it is difficult to give accurate figures of the fish catch, which has varied greatly since the lake formed.[8] The rich fish production in the earlier years of the lake's existence could have resulted from the diverse flooded substratum affecting the physical and chemical characteristics of the newly formed ecosystem, while its biological components may have contributed to the organic sediment and the fertility of the new system. More than fifty-seven species of fish belonging to sixteen families were recorded in Lake Nasser during its first decade.[9]

Figure 2.3: Camp for Nile Valley fishermen. Photograph by Dick Hoek.

This huge water body in the heart of the great Sahara desert has become increasingly important as a wintering area for migratory water birds. Records of bird species showed a huge increase from nineteen species in 1981 to forty-seven in 1990, and as many as 122 species in recent records. The number of water birds wintering on the lake could be in excess of 200,000.[10] There may be environmental hazards relating to the possible transport of disease vectors by migratory birds carrying pathogens or parasites. Various authors have recorded the impact of increasing bird populations on the physicochemical and biological water parameters (e.g., increased pH, conductivity, Biological Oxygen Demand (BOD), organic matter, nitrogen, phosphorus, coliform bacteria, and plankton).[11]

Effects of the Lake on the Desert Ecosystem

The formation of Lake Nasser has had a major impact on the ecology of the desert ecosystem in which it is located. The lake water penetrates well into the surrounding desert because of the low and relatively flat relief of the area. Wadis draining the rocky Eastern Desert are relatively deep and narrow on the eastern side of the lake, and the largest of these is Wadi Allaqi, which originates deep in the Red Sea Hills. Water entered Wadi Allaqi between 1967 and 1972 when the Aswan High Dam was built and Lake Nasser filled. About eighty kilometers of the wadi was initially inundated and remained under water for several years (Figure 2.4).

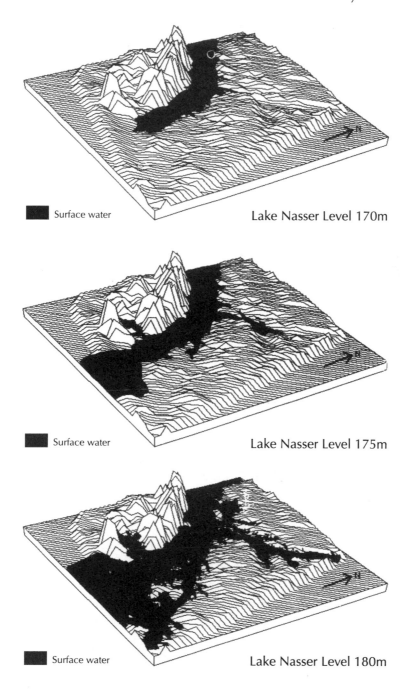

Surface water   Lake Nasser Level 170m

Surface water   Lake Nasser Level 175m

Surface water   Lake Nasser Level 180m

Figure 2.4: Topographic map of water penetration in the downstream part
of Wadi Allaqi (Selim 1990).

Prior to the lake's formation, the main source of water to support life in this extremely arid desert was occasional rainfall events. Such water accumulated in the wadi from an extensive catchment area during these rare rain events (typically occurring at intervals of many years) and was stored in the wadi-fill deposits. Since the rainfall is extremely localized and there are so few measuring stations in the desert areas, rain events cannot be accurately detected and evidence of their occurrence relies on fragmentary published evidence and direct rain records of specific areas made since the 1980s during research work in the Eastern Desert.

Occasionally, rainfall is sufficient to convert wadi floors into temporary streams for a part, or all, of their length, and there are some recorded examples of torrents caused by such episodes. Ball, writing in 1912, refers to "Linant's experience about 1830, when he recorded that the torrents from the Wadi Allaqi into the Nile were so great as to prevent his *Dahabiya* sailing up the river past the point of influx, even with good wind and all sails set." Ball also notes that a great flow of water down Wadi Allaqi also occurred in the autumn of 1902 in the Umm Gariart area in the middle part of the wadi.

More recently, observations in the Wadi Allaqi basin from 1980 to 2005 have noted fewer than ten rain events, mainly during autumn (September to November), of which five occurred, in 1982, 1986, 1994, 1995, and 1996 respectively, and all of which were sufficiently large to affect biotic parts of the system. In 1994 the torrent reached the downstream part of the wadi, at the point of influx with Lake Nasser, and discharged a considerable quantity of water into the lake. Water flow continued for ten days from 2–12 November 1994, resulting from a series of rain events in the upstream part of Wadi Allaqi. In May 1995, another torrent originating in the Red Sea Hills reached the downstream part of the wadi, and spread throughout its eastern tributaries. In the downstream part of Wadi Quleib, an eastern tributary of Wadi Allaqi, the water flowed on the surface for fifteen hours. These events show that water input is not only discontinuous but also highly stochastic. The variations in timing and magnitude of precipitation events are really quite unpredictable. Because of this unpredictability, the creation of the lake provided a new and reliable source of water for the biotic component of the ecosystem in a previously extreme arid environment.

Plentiful and reliable water encouraged the desert Bedouin to move close to the lakeshores with their livestock, using the water for human and domestic animal consumption and other household purposes. Workers from

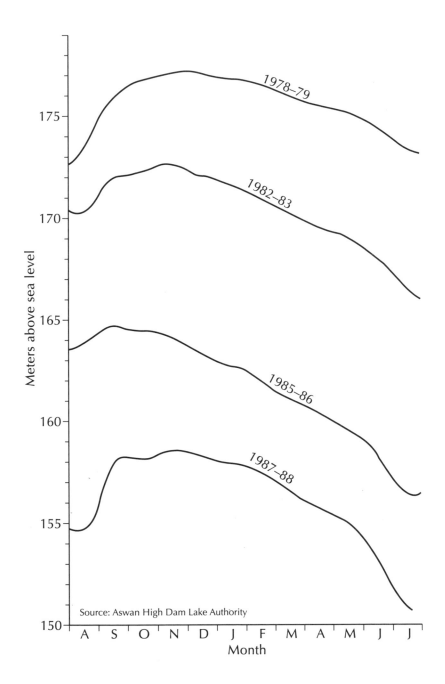

Figure 2.5: Monthly variations of lake levels in selected years.

mines and quarries in the Allaqi area also use this water, which saves them from taking long journeys to and from Aswan to bring water in water tanks. The fresh water from the lake is especially good for crop irrigation. However, commercial use of this water is not feasible because of the unpredictable fluctuations of the lake, which have frustrated attempted agricultural development on the lakeshores. A vertical fluctuation of one meter in the water level of the lake causes more than one kilometer of lateral surface-water movement, depending on the land gradient. In the downstream part of the Wadi Allaqi, a rise in the lake's water level of six meters leads to the inundation of ten kilometers of wadi floor.

The lowest water level in any year usually occurs in July and the highest in November. For example, in 1991 the absolute minimum was 162.23 meters in July and the absolute maximum was 169.35 meters in November. In 2000, the absolute minimum was 175.84 meters in July and the absolute maximum 180.63 meters in November. Despite the low water level in 1991, the seasonal variation was about seven meters, while at the higher water level in 2001 the seasonal variation was only five meters. These records illustrate the great variation in the water level, not only seasonally but also in different years (Figure 2.5).

Although used for drinking water, the shallow water of the lake is not of sufficiently good quality for this purpose. It has a high content of organic matter, especially at that time of the year when the water is rising and inundates the land surface covered with plant and animal debris, as well as human waste. This poor quality water creates problems for the Bedouin in the area, particularly the females, both women and girls, who are mainly responsible for the family's water supply; to obtain clean water they have to wade out to the deeper water and their health is affected by the cold water and ambient temperatures in winter.

On the other hand the shallow water is an excellent breeding habitat for fish, which have become abundant and easy to catch with just rudimentary equipment or even by hand. Although few Bedouin catch or eat fish on a regular basis they may buy fish from locally based fishermen (most of whom come from Upper Egypt to fish on the lake) as a food of last resort, especially when the water level rises covering any available lakeshore pasture and hence reducing goat-milk production, a main source of food for Bedouin for most of the year. In addition to providing a favorable habitat for fish, the shallow water also provides an equally favorable habitat for abundant aquatic plants, some of which are being increasingly used as feed for livestock, a feature that will be discussed in Chapter 4.

A new water source caused by the formation of the lake is water held in the soil as soil moisture after flooding. Alternating high and low water levels in the lake, producing winter inundation followed by summer exposure of the land, lead to a very dynamic redistribution of water in depressions in the wadi floor over a twelve-month period. The soil is saturated with water during the winter inundation. In summer, when land is exposed, only a small amount of the water remains in the soil, while the rest of the water is depleted as a result of evapo-transpiration and vertical seepage through the deposits.

The water held in the upper layer of soil containing plant roots is the most important source of water for plant growth. The soil moisture content in the surface layer of the wadi deposit has been monitored in eight soil profiles located at different elevations in the wadi flood area.[12] Examples of water redistribution in the soil, prior to and after the inundation, are given in the two soil profiles below both located above 170 meters ASL, where irregular flooding sequences have taken place. This area was temporarily inundated during the first lake-level peak in the late 1970s. With decreasing water levels during the 1980s the area was exposed and the arid conditions returned, where only runoff water from infrequent rain events contributed to the soil moisture in this part of the wadi. When the monitoring of the moisture content in the at these two sites soil began in 1992, the area had already been exposed for ten years. With the second rise of the lake's water level, which began in 1993, the inundation was repeated but with short intervals of land exposure in summer.

The water redistribution in the soil profile at the lower relief position, located in the wadi runnel at an elevation of 174.4 meters ASL, is shown in Figure 2.6. The soil contained only a small amount of water (1 percent of its volume) up to a depth of 135 cm, prior to the area's inundation in September 1994. The extremely small temporal and spatial variations of the moisture in the surface layers (up to a depth of 1.5 meters) of the wadi deposits indicate typical desert conditions. The soil became saturated with water during a five-month inundation period (October 1993–February 1994) and in the following exposure period (March–August 1994), while a small amount of water (about 10 percent of soil volume) was held in its topsoil layers up to a depth of one meter during the three months from March to May. Below the depth of one meter, a considerable amount of water (35–40 percent of the soil volume) was stored in the soil yet, similar to the topsoil, it was present only for the first three months following inundation. After the second, and most prolonged inundation period of ten months, the land was exposed for

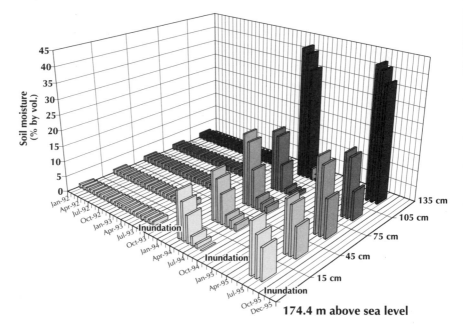

Figure 2.6: Soil moisture content at 174.4 m above sea level.

three months in summer. The amount of water held in the soil profile was similar to that in the previous exposure period. Most probably, the water rapidly decreases from the upper one meter of soil through water absorption through plant roots (transpiration water loss), evaporation from the soil surface, and water percolation through the soft deposits.

The second site for monitoring soil moisture was located on a slightly elevated wadi terrace at a height of 176.2 meters ASL. Because it is situated two meters higher than the previous site, it was inundated a year later (October 1994), followed by six months of exposure in the spring–summer period (April–September, 1995). The amount of water held in the soil prior to inundation and to the redistribution of water through the soil profile during the exposure period was similar to that in the lower gradient shown above.

These two examples of water redistribution in the soil profiles provide evidence that the alternating periods of flooding and exposure have a major effect on the amount of water stored in the wadi sediments. The frequency of floods and their duration depend on the wadi's topography, while the rate of the vertical infiltration of water is affected by the coarseness of the wadi sediments.

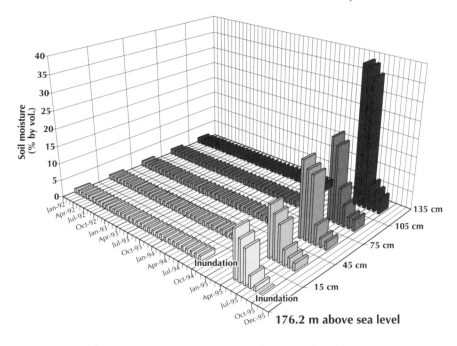

Figure 2.7: Soil moisture content at 176.2m above sea level.

Because the growth of the shallow-rooted plants cultivated for human consumption depends on the availability of water in the surface soil layers, knowledge of water availability is very important for the development of this periodically inundated land. Two extreme scenarios are possible in the flood area: water deficit between the sequence of floods and water excess during the inundation period. As illustrated above, topsoil to a depth of one meter lost water, and dried up, in a period of time considerably shorter than even the fastest-growing crops could reach maturity and be harvested. On the other hand, on the frequently inundated land the period of land exposure is much shorter than the life span of cultivated plants. It seems, therefore, that floodwater stored in the wadi sediments cannot be used for sustainable agricultural purposes (shallow-rooted crop cultivation) without irrigation. An alternative development, however, is to grow deeply rooted drought- and flood-resistant plants (trees and shrubs), so using the flood land for agro-forestry.

The main sources of underground water in the Wadi Allaqi basin are both rainfall water originating in the Red Sea Hills and water infiltrating from the lake in the downstream part of the wadi.[13] In the area outside the lake's influence, rainfall water is the only source for recharging the subsurface water. After rain, the surface runoff water is drained from the extensive basement complex area to the lowest areas of the desert landscape. This water is stored in the fissures and faults of the rocks and in wadi deposits. Torrents or surface runoff water rarely originate in the Nubian sandstone area, which is covered by sand sheet and where the wadi drainage system is poorly developed. Rainwater favors only local areas, while the rest of the terrain remains dry and water can be found only in the area in contact with the water body.

Wadi sediment is a potential aquifer in the Wadi Allaqi area. The groundwater table in the main channel of the wadi, beyond the influence of the lake, is at a depth of between twelve to twenty-three meters below the wadi surface. Deeply rooted trees and shrubs growing in the main wadi channel and its tributaries provide good evidence of the presence of underground water or a wet deposit layer. Roots of acacia, which dominate the plant communities in the wadis of the Eastern Desert, penetrate the wadi deposits, sometimes to a depth of twenty to thirty meters. Water stored in the deep deposits is the only source of water that supports their growth during the prolonged drought periods, which may extend for several years.

Deeply rooted trees and shrubs growing in the wadi play an important role in the hydrological cycle through transpiration, and hence the redistribution of moisture in wadi-fill deposits. Water is lost from the soil and returned to the atmosphere through evapo-transpiration. Evaporation of water from plant leaves causes suction in the plant and its roots, so that the water held in the soil moves to the plant roots. Redistribution of soil water from horizons deeper in the profile to dry surface horizons by the root system is termed 'hydraulic lift.' This process explains the high stability of desert trees; once successfully established, they grow for long periods of time if not disturbed either by humans (most often the cutting of trees) or by natural hazards such as being torn down by wind or uprooted by strong torrents.

The existence of vegetation in the wadi channel helps to preserve water, encouraging the deposition of soft alluvial material by slowing down the velocity of runoff water (torrents) and therefore increasing water infiltration downward into the wadi deposits. Plant roots anchor the soil, preventing erosion, while aerial parts of plants decrease water evaporation from the soil surface by shading the ground. The position of a tree in the wadi is a good indicator of the presence of underground water. Studies conducted in the Eastern Desert have revealed that the topography of the wadi has a great effect on water availability, and hence the distribution, of trees and shrubs.[14] Narrow wadis covered by shallow deposits constitute the most favorable habitat for the deeply rooted vegetation. Despite more water being stored in the deeper, rather than in the more shallow deposits, water can quickly penetrate the deposits below the root zone, thus quickly becoming unavailable to the plants. On the other hand, in wadis covered by shallow, coarse deposits, rock cracks allow and facilitate water passage to the roots. Thus seemingly shallow soils may, in fact, be ecologically quite suitable.[15]

The exploration of the underground water in the wadi sediment is not an easy task. During the implementation of the project on the cultivation of medicinal plants (1993–1995) the REGWA Company constructed three wells in Wadi Allaqi, but in areas beyond the influence of the lake water. No water was found in those parts of the wadi covered by shallow deposits (12–18 meters) that were underlain by basement rocks. Interestingly, the Bedouin of Wadi Allaqi have experience in finding underground water through observations of the wadi's topography and plant growth, and they locate and dig their wells accordingly. Often they select areas for wells on the wadi terrace close to trees. Water in such wells is typically found at depths of four to ten meters; its yield, however, is invariably very low, varying between four and twenty cubic meters a day, and, thus, this water tends to be used only for domestic purposes.

Prior to the construction of the Aswan High Dam and the lake formation there were many shallow wells in the wadi. These wells were used not only by the local Bedouin in Wadi Allaqi, but also by travelers passing through on their way to Sudan. A story about the well in Wadi Umm Ashira at its connecting point with Wadi Allaqi, which is about sixty kilometers from the Nile, deserves mention here. A Russian archaeological expedition that studied Wadi Allaqi in the 1960s prior to inundation by the lake discovered that an inscription on a *stela* found at the Kuban fortress referred to the Umm Ashira well,[16] which was dug during the third year of the rule of Ramesses II and was 6.24 meters deep with two meters' depth of water. This site was the most important stop for donkey caravans en route to the gold mines. In the late nineteenth century there was a military camp here to prevent the Mahdi movement entering Egypt from Sudan, and the British soldiers dug a well in the same location but did not find any water.[17] However, since the lake formation there has been water in the Umm Ashira well, which is used by the local population and for the irrigation of the South Valley University experimental farm.

Water is stored in the fissures and faults of rocks at depths well below that of the wadi floor. Such water was used in the past by miners for both domestic purposes and for crushing ore to extract gold. Many gold mines of different sizes functioned in Wadi Allaqi from as far back as the pharaonic age until the beginning of the nineteenth century. The gold mining shafts of Umm Qareiyat and Abu Sweyel, dug in the basement rock, have since filled with underground water. The depth of the water level in these shafts is twenty-five meters below the surface,[18] while water is also found at a depth of fifteen meters below the wadi surface in the nearby Wadi Haimur well.[19]

In the downstream part of the Wadi Allaqi, which is periodically inundated by lake water, the groundwater in the wadi sediment comes from two sources: surface runoff from rain in the upstream part of the wadi and water infiltration following inundation by the lake. In the flood plain, the recharge caused by vertical infiltration is more important than the horizontal groundwater flow to the upstream part of the wadi, which is very minor and restricted to a relatively small distance from the lakeshore of two kilometers in winter.[20]

In the wadi flood plain the sediments are saturated during the inundation period. After the lake recedes, exposing the area that had been flooded, water infiltrates vertically and recharges the groundwater. The time interval between any two floods is the main factor affecting the recharge of the groundwater. The long-rooted tamarisk shrubs are a good indicator of groundwater in arid climates; their rich growth in the downstream part of Wadi Allaqi demonstrates the presence of water in the wadi sediments of the flood plain. De Vries and van Zanten (1993) found that the groundwater table in the downstream part of Wadi Allaqi is between ten and fifteen meters deep. Six piezometers, which were installed in the part of the wadi that had been exposed for almost thirteen years, showed the sediment's aquifer to be at a depth of from eight to twelve meters.[21] Water in this aquifer is directly affected by the water level of the lake as it varies throughout the year, the aquifer containing more water at those times when the level in the lake is higher, and vice versa.[22]

In the frequently inundated part of the wadi, the groundwater is close to the surface. This is the area where numerous shallow wells are dug by local people to obtain water for domestic use and for the irrigation of small cultivated plots. These Bedouin say that water up to a depth of ten meters is still available three years[23] after an inundation, while the water level in shallow wells drops below two meters in three months.[24] The seasonal and long-term variations of water quality in these shallow wells depend on the

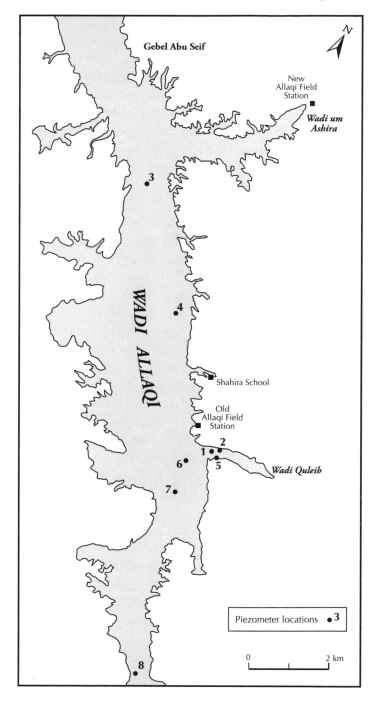

Figure 2.8: Location of piezometers in Wadi Allaqi.

water level of the lake, the depth of the well, its distance from the lake-shore, and the nature of the water bearing substrata (wadi deposits).[25] In locations close to the lake shoreline during winter, when the water level of the lake is high, the safe yield for wells is fifteen to a hundred cubic meters per day. However, at the same locations during summer, when the water retreats, the level of the water in the wells drops considerably, giving a safe yield in a range of seven to forty cubic meters per day.[26] This is the time of year when the Bedouin move closer to the lake and begin digging new shallow wells near the shoreline, with a depth of up to three meters. Water usually remains in such wells for up to two to three months prior to the annual autumnal inundation.

The chemical composition of the underground water varies, depending on the location of the wells. The most important characteristic of the water quality is the amount of total dissolved salts, which is shown by the electrical conductivity (EC) of the water samples.[27] The water chemistry in the shallow wells located close to the shoreline is similar to that of the lake water, which is about 0.2 dS cm^{-1}.[28] At a distance of about fifty meters from the shore, the electrical conductivity (EC) of water samples is about 0.6 dS cm^{-1}, the pH is less than 7.5, and the concentration of soluble salts is low.[29] However, water in such wells is of better quality for human consumption than the lake water because the water has been filtered by passing through the sandy deposits and hence becomes less rich in organic compounds than the surface water. Yet, after filtering, the concentration of soluble salts in the underground water rises with increasing distance from the lake. The electrical conductivity of the water sample taken from the piezometer located close to the water edge (piezometer 3 in Figure 2.8) is much less (0.47dS cm^{-1}) than those situated further from the lake, where water is slightly saline (piezometers 4 and 7 with ECs of 0.83dS cm^{-1} and 0.75 dS cm^{-1} respectively). The highest EC, together with the highest concentration of salts, was observed in piezometers located outside the flood areas of Wadi Allaqi and its tributary Wadi Quleib (piezometers 2 and 8 with ECs of 1.64 dS cm^{-1} and 4.08 dS cm^{-1} respectively).

The distance from the lakeshore, however, is not the only criterion determining the quality of the underground water. Wadi topography, which affects the redistribution of water and wadi sediments, plays a significant role in both the quality and quantity of the groundwater. In Wadi Allaqi, at the meeting point with its tributary Wadi Quleib (piezometer 6), the underground water is sweet, with a low salt content and a low EC of 0.63 dS cm^{-1}. Recharge of the underground water in this part of the wadi is enhanced by runoff water from Wadi Allaqi and Wadi Quleib, and from the vertical infiltration of the lake water, when this area is inundated.

The likelihood of the presence of sweet water in the wells is low, even when new wells are dug in the same area as those wells already containing sweet water. The water chemistry in the wells varies considerably and is strongly influenced by the salt content found in the water-holding substrata. Water from only one of the three wells dug close to each other in Wadi Allaqi at the Egyptian Environmental Affairs Agency field station is suitable for both domestic use and the irrigation of cultivated plots.

The concentration of the soluble salts in a well will be higher if the well is not closed, as a result of high rates of water evaporation from the open water surface in the well, leading to an accumulation of salts. Frequent withdrawal of water from a well tends to improve its quality.

The soils and sediments in Wadi Allaqi largely consist of wadi-fill deposits that vary in depth, in both physical and chemical composition, depending on the soil-forming materials, and on transport processes.[30] The main channels of Wadi Allaqi and its tributaries are covered by deposits that are an accumulation of rock materials of different grain sizes. This material is transported by runoff water as alluvial deposits from the upstream part of the wadi, and by wind transport as aeolian deposits from the surrounding desert sand sheets. The running water carries suspended particles of rock, gravel, sand, silt, clay and even plant parts, which gradually settle out where the wadi widens and the runoff current slows down. Trees and shrubs growing in the wadi also trap particles that are suspended in the runoff water. These form small hillocks on the side of plants facing downstream, some up to one meter in height. The alluvial sediments in the wadi channel vary in texture from relatively coarse pebbles and rock fragments to fine silt and clay particles arranged in horizontal layers of different thicknesses ranging from one to three centimeters, reflecting the difference in intensity of runoff events.[31] The layering pattern is clearly observed in the dry wells located in Wadi Allaqi; these layers are distinguished by their color, texture, and the presence of shell and fossil roots.[32] Softer deposits in the surface layer, mainly silt and clay, overlay the coarse deposits as a result of the redistribution of sediments by the runoff water. This layer conserves water by preventing its evaporation from the subsurface deposits and creates a seed bank by trapping the seeds and providing favorable conditions for seed germination. The thickness of such a surface layer varies from one to ten centimeters; it may remain for more than a year, or disappear through wind erosion within only a few weeks. Severe erosion by wind particularly occurs

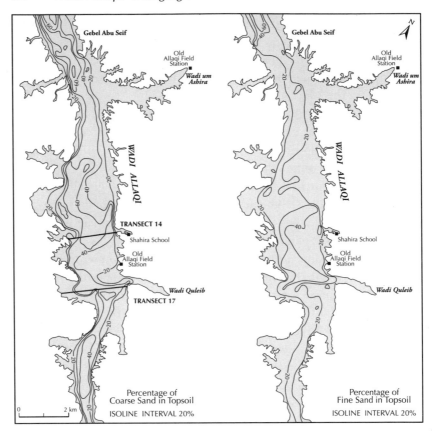

Figure 2.9: Distribution of coarse sand and fine sand deposits in the topsoil in Wadi Allaqi.

where the surface crust is disturbed either by vehicles passing through the wadis or by camel trains.

Small tributaries with steeper gradients are characterized by a larger proportion of coarse deposits such as rock fragments and pebbles, while the sediments in the main Wadi Allaqi channel tend to be much finer than in the tributaries. Even finer-grained material tends to be deposited along the edges of the wadi, while coarser material is found in the centre of the wadi in the runnels, reflecting the transportation ability of the faster-flowing water in the runnels themselves when flow episodes occur.

The thickness of the wadi sediment varies considerably depending on the local topography, gradients found in the wadi, and the size of the catchment area. The maximum thickness of sediment for the main channel of Wadi Allaqi is about sixty meters, and some forty meters in the main tributaries.[33]

In the middle part of Wadi Allaqi the old silt terraces have been subjected to heavy erosion, but are still recognizable in a few locations. The fossilized hillocks of a previous woody cover are well preserved in this part of the wadi. It is likely that the majority of waterborne deposits were laid down during the wetter period in the Holocene, about 5000 BC.[34] In the present dry conditions, the rare flood events make only a minor contribution in the accumulation of deposits on the wadi floor. An additional minor process contributing to the wadi sediments is the deposition of wind-blown material (aeolian deposits) into soft deposits of orange sand that collect mainly at the edge of the wadi. The distribution of coarse- and fine-sand fractions suggests cross-wadi variation.

The soils on the flood plain in the downstream part of Wadi Allaqi are modified by material left by receding lake water, in addition to that deposited by runoff water and wind. The deposition of soil material from runoff water following rain events is similar for all wadis, with mainly the finest material mainly reaching the downstream part of the wadis. Fan-shaped deposits of fine sand and silt are found on the side of the larger plants facing downstream, notably *Pulicaria crispa*, which is a compact shrub and has a significant effect on the rate of the water flow.[35]

The most significant factor influencing soil quality in this part of the wadi is that of the lake water. There are two important processes taking place: the deposition of silt from the lake during inundation; and changes in the chemistry of the surface soil layers during, and immediately following, inundation. Flooding by lake water results in the deposition of lacustrine material, usually consisting of silt, shells, and organic matter, on the soil surface in those parts of Wadi Allaqi subjected to inundation.[36] The depth of the sediment deposited by the lake water is shallowest, about two centimeters, in that part of the wadi that is only briefly inundated. In those parts of the wadi subjected to longer and more frequent periods of inundation, deeper layers, sometimes up to ten centimeters, of lake-deposited sediment are formed.[37]

Alternating periods of flooding and exposure have resulted in chemical and microbiological changes in the flooded soil. On the recently inundated soils in the areas closest to the lakeshore, a thin crust of up to one centimeter, orange or orange-brown in color, covers the soil. This is underlain by a dark gray-black layer that is thin (five to ten centimeters) and rich in sulfide. This layer eventually disappears with increasing distance from the water's edge.[38] A combination of these two layers indicates that flooding of the soil by lake water has resulted in a reduction in the surface layers. Oxidation-reduction (or redox) changes in the soil surface have a significant influence on the chemistry of essential trace metals (especially iron, manganese, copper, and zinc) and their availability to plants.[39] The alternating reducing and oxidizing conditions result in concentrations of iron and manganese by precipitation of these oxides.[40] Further away from the lakeshore, in soils that have been exposed for longer periods of time, a light gray or grayish brown surface horizon of three to twelve centimeters in depth is formed. This color may result from the longer period of weathering under oxidizing conditions.[41] Because the subsoil in these profiles is fairly uniform, being red to reddish-brown in color, major changes due to flooding may take place in the upper part of the soil profile.

A soil survey conducted in the Wadi Allaqi flood plain shows large variations in key soil parameters (Table 2.2). This may be owing to various processes acting on the soil, particularly the sorting by size of soil particles by runoff water in occasional torrents, mainly in the past, and by lake water. The pH range shows an alkaline reaction that is usual for soils in Eastern Desert wadis. A broad range of electrical conductivity (EC) values are shown in the table, but severe salinity, with a value above 3 dS cm^{-1}, was recorded in only 2 percent of all 212 soil samples. In 66 percent of the samples the EC, has less than 0.7 dS cm^{-1}, indicating non-saline conditions similar to those recorded throughout Eastern Desert wadis.

Organic matter is an important fraction of the soil as it has an important role to play in nutrient retention and cycling, water relations, and structural stability. Its presence in the soil, estimated by values of loss on ignition (LOI), is generally low except in some limited areas where litter has accumulated under a dense growth of tamarisk. These low LOI values point to the small amounts of organic matter present in the soils. Extractable calcium, manganese, potassium, and sodium are the four macronutrient cations providing a measure of the soil's nutrient-supplying ability; their concentrations in the soils of the flood plain are sufficient for crop production.[42] Distribution of some, but not all, metals seems to be affected by the lake water. Correlation analyses indicate that the highest concentration of sodium occurs in recently inundated soils close to the lakeshores, while the distribution of calcium differs in that its concentration increases with distance from the lake.[43] Use of acid oxalate, an extractant that removes oxides and associated trace metals, has indicated the presence of significant amounts of iron and manganese oxides, both of which are known to absorb and occlude trace elements such as zinc and copper, essential for crop growth.[44] A comparison of key soil parameters, particularly pH, EC, and LOI, in the downstream part of Wadi Allaqi made prior to inundation by Kassas and Girgis (1970) with these post-inundation parameters indicates that the soils have undergone relatively little change since then.[45]

Table 2.2: Summary of data from survey of the soil in Wadi Allaqi flood plain (1988–1989).

Soil properties	Range of values
pH	7.0–9.1
EC	0.24–7.06 dS cm^{-1}
LOI	0.1–7.2%
Ca*	3600–11700 mg kg^{-1}
Mg*	136–589 mg kg^{-1}
K*	102–570 mg kg^{-1}
Na*	115–666 mg kg^{-1}
P**	8.9–96.2 mg kg^{-1}
Fe***	204–2007 mg kg^{-1}
Mn***	52–526 mg kg^{-1}
Cu***	3.0–10.8 mg kg^{-1}
Zn***	1.8–8.0 mg kg^{-1}
Al***	151–1067 mg kg^{-1}

Note: Data presented in this table is the summary of the soil survey in 1988–1989 when the lake was at its lower layer (154.3 meters ASL). 24 transects were set across the wadi along the distance of about 15 km of flood plain. Soil samples (from both topsoil 0--10 cm and subsoil 10–50cm) were collected from the quadrates located at 100 meters interval along the each transect.

Sources: Moalla and Pulford (1993a) (pH, EC, LOI), Moalla unpublished (P) and Ali, Badri, Moalla, and Pulford 2001 (Ca, Na, K, Mg, Cu, Zn, Al, Mn, Fe). Ammonium acetate extractable*: Ca, Mg, K and Na (1 M ammonium acetate pH 7); Sodium bicarbonate extractable**: P; Oxalate extractable***: Cu, Zn, Al, Mn and Fe (0.175 m ammonium oxalate 0.1 M oxalic acid pH 3.3^{-1}).

Soils in the flood areas are generally non-saline, except on the lakeshore itself and in those areas colonized by tamarisk. The large soluble salt content, mainly sulfates, in the surface soil layer is the result of the tamarisk's ability to recycle salts. The type of soil in the downstream part of Wadi Allaqi is close to Entisols sub-order Psamments in the soil classification scheme given by the United States Department of Agriculture (USDA).[46]

Vegetation of Wadi Allaqi

Rain remains the single source of water for the biological components of the ecosystem in the areas outside the lake's influence. Plant growth in the Wadi Allaqi area is restricted to the wadi channel where runoff water collects, recharges the groundwater, and is stored in the wadi sediments. This water is sufficient to maintain the water balance of xerophytes (plants with special adaptations to an arid environment) during the prolonged rainless periods that may extend for several years. The vegetation of the upstream part of Wadi Allaqi and its tributaries, beyond the lake's influence, comprises two types: permanent vegetation dependent on accessible groundwater, and opportunistic vegetation dependent on unreliable and unpredictable rain water stored in the upper layer of wadi deposits.

Permanent vegetation consists of drought-resistant perennials, primarily deeply rooted trees and shrubs, forming a permanent framework of plant cover on the wadi floor. These plants make economical use of limited moisture by developing extensive root systems and maintaining wide spacing between individuals. In the middle part of Wadi Allaqi, no more than one mature individual of acacia (*Acacia tortilis* subsp. *raddiana* and *A. ehrenbergiana*) is established within an area of ten hectares. In the upstream part of Wadi Allaqi, where the woody growth is subject to little human disturbance, about seven trees and shrubs of different species (acacias, *Balanites aegyptiaca, Ziziphus spina-christi, Salvadora persica*) normally grow in an area of about one hectare.[47] The most common trees in the wadi are acacia and *Balanites aegyptiaca*. Both trees have two root types: roots that lie close to the soil surface, often being partly exposed, and tap roots that penetrate wadi deposits to depths of up to twenty meters. Numerous surface roots, each of which could be more than fifteen to twenty meters in length, create a network that collects and stores water from the surface layer after occasional rain events. Experiments conducted in Wadi Allaqi in February 1994, three months after rain, showed that the surface roots of *Acacia tortilis* subsp. *raddiana* contained a large amount of water; a five-meter piece of the surface root with a diameter of about one centimeter contained an average of about twelve milliliters of water.

In dry periods, the growth of acacia is supported by deep roots that tap the underground, subsurface water. During the long rainless periods when underground water is not recharged, many trees and shrubs, such as acacia, shed their leaves, resulting in a reduction of water loss through transpiration. Reduction of the transpiration surface by the development of minute leaves, for example, tamarisk, or even leaflessness (*Capparis decidua* and *Leptadenia*

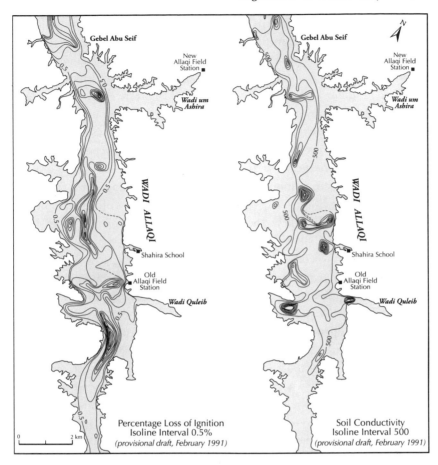

Figure 2.10: Distribution of loss on ignition and soil conductivity in Wadi Allaqi.

pyrotechnica), are among the other important adaptations to water economy in arid conditions.[48]

The availability of groundwater in the upstream part of Wadi Allaqi supports some permanent vegetation dominated by trees and shrubs. Permanent vegetation types are the key component of the desert ecosystem. Deeply rooted perennials, which dominate this vegetation type, are the only ecosystem producers supplying organic matter to consumers, including wild animals and the livestock of local Bedouin, during prolonged rainless periods. These plants play an important role in the local economic system by providing permanent grazing facilities, fuel, charcoal, and many other benefits to help people survive in this area.

In certain localities in the main wadi channel the plant growth acquires a form that may be described as desert woodland. This contains trees of *Acacia tortilis* subsp. *tortilis, Acacia tortilis* subsp. *raddiana, Balanites aegyptiaca,* and *Ziziphus spina-christi,* which may be covered by lianas, such as *Cocculus pendulus* and *Ochradenus baccatus.* Evergreen shrubs (*Salvadora persica, Leptadenia pyrotechnica,* and *Ochradenus baccatus*) can experience vigorous growth, reaching the height of small trees in some cases. Some trees and shrubs even grow on the wadi sides in the cracks between rocks where pockets of underground water are available.

The floristic composition of phanerophytes (trees and shrubs) in the upstream part of Wadi Allaqi is closely related to that of phanerophytes growing in the Nile Valley. The First Cataract Islands on the River Nile near Aswan, where Nile Valley relict vegetation is conserved, contain some trees in common with the upstream part of Wadi Allaqi, particularly *Acacia tortilis* subsp. *raddiana, Faidherbia albida* (better known by its old name *Acacia albida*), *Balanites aegyptiaca, Ricinus communis, Ziziphus spina-christi,* and *Capparis decidua.* The presence of these plants, together with *Tamarix nilotica,* in the upstream tributaries of the Wadi Allaqi indicates edaphically favorable moisture conditions in this extreme arid region.

The middle part of the wadi is characterized by deep sediments and supports a geographically wide distribution of acacia trees (*Acacia tortilis* subsp. *raddiana* and *Acacia ehrenbergiana*). Patches of *Salsola imbricata* shrubs are associated with the mouths of tributary wadis. Numerous fossil hillocks occurring in the middle and driest part of Wadi Allaqi with dead remnants of *Tamarix aphylla* and *Salvadora persica* (carbon-dated to 500–800 years ago) appear to be relics of extensive *Tamarix* and *Salvadora* thickets.[49] This indicates that more pluvial conditions may have been prevailing in the recent past. Many kilometers of the wadi floor now have no living plants, but just the dry remains of vegetation from previous wetter periods when rainfall was more abundant.

After rain, runoff water moistens the deposits of sediment in the wadi and activates various biological processes, in particular the germination of seeds that have been lying dormant in the soil seed bank. Water availability, intensive solar radiation, and high air temperatures simultaneously trigger photosynthesis, and plants develop a relatively large biomass in a very short period of time of some two to three months. For example, the phytomass of the *Cullen plicata* community in the middle part of Wadi Allaqi after the rain event in October 1994 was estimated to be as high as 136.5 grams per square meter of air-dry weight, or 1365 kg/ha^{-1}. In comparison, the dry matter production of this desert plant is similar to the dry matter yield of the leguminous *Medicago* sp. (dry yield of 1200 kg/ha[1]), and much more than the mixture of *Trifolium repens* (dry yield 250 kg/ha[-1]), which is cultivated for forage in tropical countries.[50]

This is opportunistic vegetation, comprising annuals and short-living perennials that are linked to immediate water availability.[51] The life cycle of these plants depends entirely on the amount of water stored in the ground. These are inherently perennials that have the ability, under stress of water shortage, to complete their life cycle within a number of months (or maybe a few years), and to produce seeds that remain in the dormant stage until the next rain event. In the extreme arid conditions prevailing in the Eastern Desert, most of the potential annuals are able to produce seeds in a period of six to ten months and complete their life cycles within one year. Dwarf plants (*Senna alexandrina*, *Fagonia* sp., *Cleome droserifolia*, *Cullen plicata*, and many others) with few, but well-developed fruits are very common. In long rainless periods most dry plants are blown away by wind action; those that remain in the ground resume growing and are able to produce new offspring, even with a light shower that is not sufficient for seed germination.

In spite of its irregular appearance and short life span, this opportunistic vegetation, by providing rich pasture resources, is a key driving force behind the Bedouin way of life. Bedouin move their livestock, almost exclusively sheep and goats, across the desert, following rain and searching for pasture. Even those Bedouin communities that are settled in rural or urban areas still frequently send their livestock to the desert to graze on ephemeral pastures. For those who are semi-settled in the downstream part of the wadi close to the lake, sending livestock for grazing on such pastures has remained an essential component of Bedouin livelihood. This will be discussed at greater length in Chapter 4.

Ecotone in Downstream Wadi Allaqi

An ecotone system in the downstream part of the wadi formed in the lake-desert transitional zone in the 1960s. Fluctuations in the lake water level (Figure 2.2), which can be as much as thirty meters vertically in the medium-term, with a horizontal expansion of about fifty to seventy kilometers, control the dynamics of the ecotone system. Ecotonal systems differ from zonal ecosystems because the life strategy of ecotonal biota must provide a stable existence in an unstable environment. Such environments are characterized by a higher frequency of, and a wider range of, fluctuations of their key characteristics; of these, water is the major controller of the desert ecotone because of its extreme differences between dry periods and flooding events.

The response of the system to a regular sequence of flooding and exposure periods depends on the time interval between events. Exposure of the land, and hence available oxygen, activates biological processes and the build-up of plant biomass (annuals) or intensively increased growth (perennials). Many ecological niches, with different moisture regimes, represent the unique property of the desert ecotone.

Growth of annuals is restricted only to the short period following inundation. Woody perennials, such as tamarisk, continue growth until water in the wadi-fill deposits becomes limited, at which point the growth of plants slows down. Flooding stops biological activity, and hence biomass production, and with long inundation periods the biomass decreases considerably. An important difference between the ecotone and the desert ecosystem is that there is no seed reserve available in the ecotone. Seeds are removed by the floodwater, and new seeds arrive from neighboring areas, transported by water or wind.

In the extreme conditions prevailing on the shores of Lake Nasser, the ecotone comprises a low biodiversity, while bioproduction is high. Thirty angiosperm plant species have been recorded, among which only five species are abundant. *Tamarix nilotica* (tamarisk) is dominant, providing a high phytomass (average of 26.95 kilograms per meter).[52] Tamarisk tolerates a wide ecological amplitude, and it can survive, and even vigorously grow, during an interval of more than fifteen years' inundation, as well as withstanding complete inundation for a few years. Water-spenders, such as *Tamarix nilotica*, that lose water at high rate as long as it is available, have a special adaptation to excess water during the short exposure period and to water limitation during the long exposure period. Adaptations for water spending include a high ratio of conducting to non-conducting tissue; high root to shoot ratio; high water absorbing potential; and the ability to become a water saver under extreme stress. Once it has colonized an area, tamarisk becomes a strong edificator in plant communities by changing the environment, particularly by shading the soil surface and increasing the soil salinity of the upper layer through salt recycling, thus preventing the invasion of new species.

The vegetation pattern of the flood plain, or ecotone, is organized into zones according to water availability and length of flooding periods. Shallow water close to the lakeshore sustains a high density of phytoplankton, while water blooms, caused by flourishing *Cyanophyceae* species, occasionally occur. Among the submerged euhydrophytes in shallow water the most common are *Najas* spp., *Ceratophyllum demersum, Vallisneria spiralis,* and *Potamogeton crispus*. The frequently inundated terrestrial zone close to the water's edge is covered by a compact layer of dry aquatic plants (mainly *Najas* sp., which dominates aquatic plant communities in the shallow water). Bedouin children often collect dry plants, rolling them into balls and feeding them to animals when no other food is available. The numerous small circular depressions (about one meter in diameter and up to fifty centimeters deep) that are produced by the nesting activities of fish act as retaining pools of water. In this area, the groundwater is very close to the soil surface (ten to fifty centimeters). Growth of annual herbs is restricted to this short period while water moisture is available. They include *Glinus lotoides* var. *dictamnoides, Heliotropium supinum, Portulaca oleracea* and *Amaranthus blitoides,* grasses (mainly *Cynodon dactylon*), and sedges (*Crypsis schoenoides, Cyperus michelianus* subsp. *pygmaeus,* and *Fimbristylis bisumbellata*). Many of these plants survive only to the seedling or young plant stage; the small plants are easily destroyed by submergence through the fluctuating water level in this zone, often before completing their life cycle. Fast drying of the surface soil layer is another reason for the short life cycle of the plants. Intensive grazing in this zone contributes to the sparse plant cover and trampling causes soil erosion, once the soil dries out.

Adjacent to the frequently inundated zone is an annually inundated transitional zone at some distance from the water's edge. The groundwater in this zone varies from one to four meters below the soil surface. Conspicuous here is the recent deposition of lake sediments, comprising alternating layers of coarse and fine particles; the soil is characterized by a fairly low content of total soluble salts. Vegetation is dominated by a vigorous growth of *Glinus lotoides,* which thickly carpets the ground. There are numerous seedlings and young tamarisk, together with old shrubs that have survived inundation. Shrubs of tamarisk become predominant with increasing distance from the lake, while the growth of *Glinus* declines under the dense canopy of tamarisk.

The third zone has a history of gradually reducing partial submergence with increasing distance from the lake. It is not annually inundated and the growth of plants is supported by the water from previous inundations stored in deposits at a considerable depth (three to six meters), or by seepage (horizontal water movement) from the lake. Recent deposits accumulated under tamarisk contain a high proportion of fine silt. The vegetation is dominated

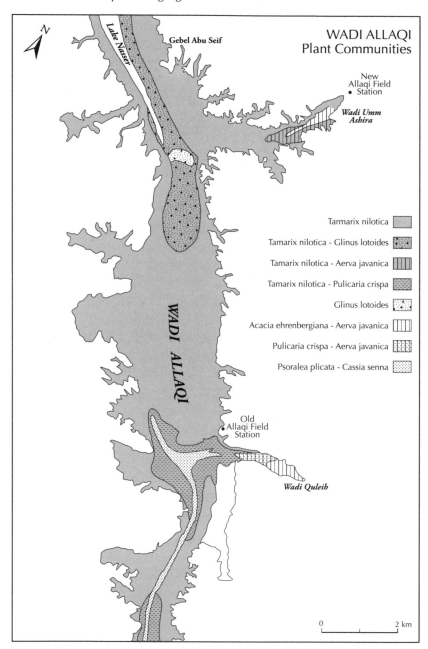

Figure 2.11: Plant communities in Wadi Allaqi.

by tamarisk, which has the important function of recycling salts: the roots absorb the salts, which are concentrated in the leaves. These salts are exuded by salt glands on the leaf surface and eventually result in the death of leaves, which fall and accumulate on the soil surface. In the extreme, arid ecosystem prevailing in the southern part of the Eastern Desert, where it very rarely rains, the litter of tamarisk does not decompose but is gradually removed by wind action and does not contribute to soil salinity.

The situation differs in the flood plain, where dense thickets of tamarisk prevent the litter from being blown out by wind and a huge amount of litter accumulates under the shrubs, up to one meter in height in places, partly covering their low branches. Because humidity is higher in close proximity to the large water body, some decomposition takes place and, when the litter is partly decomposed, it adds salts to the soil. The dense growth of tamarisk forms an almost mono-species zone in the middle part of the wadi with the height of shrubs reaching seven to eight meters. Tamarisk is so tightly packed that it is very difficult to pass through without cutting branches. Each shrub usually has from three to six main branches, growing from the base of the shrub. In some areas, the dense tamarisk can prevent access to the lakeshore. Workers from the nearby mines have burned and bulldozed patches of tamarisk in order to gain access to the shore to obtain water from the lake.

Further away from the lake another zone has appeared, making a vegetation transition where desert elements have come to play significant roles in the plant structure. Here, *Pulicaria crispa* is a co-dominant species in the tamarisk community. The effect of the lake diminishes with the distance from its shore, but a fairly thin layer of lake deposit, containing a high proportion of fine silt, provides evidence of past inundation. Soils here have a fairly low total soluble salt content, decreasing steadily with the depth of soil and distance from the lake. Groundwater, which is a result of past inundation, is found at a depth of seven to ten meters. With increasing distance from the lake, tamarisk growth becomes sparse while *Pulicaria* takes over and increases in number.

The last zone on the edge of the flood plain where the effect of the lake is still noticeable is dominated by *Pulicaria crispa*, with a considerable occurrence of typical desert plants. The soil consists of typical desert wadi deposits comprising alternating layers of fine and coarse materials. The effects of past floods remain noticeable through the presence of circular depressions left by fish nests and remains of silt brought by lake water. Individual stands of *Pulicaria* here form a huge, almost spherical outgrowth up to two meters high and four meters across. The distribution of the desert plants growing in this zone is restricted to the lower portion of the wadi floor, particularly to the main runnels where the plant community is dominated by *Cullen plicata, Aerva javanica,* and *Haplophyllum tuberculatum.* Other plants characteristic of the desert wadis are *Senna alexandrina, Convolvulus deserti,* and *Citrullus colocynthis.* Most probably the seeds of these plants are brought there by runoff water from the upper part of the wadi after rain. The new *Pulicaria-Cullen* community becomes noticeable at the end of the flood plain on its sharp border with the desert plain, which has sparse vegetation cover in the rainless period. *Cullen plicata* dominates the typical desert communities after the rain events, providing excellent pasture for livestock in Wadi Allaqi.

This last zone provides a suitable habitat for desert trees, the seeds of which are occasionally brought by the runoff water from the upstream part of the wadi, as well as by humans and their livestock. Runoff water, which moistens the surface layer of the wadi deposits, activates the germination of seeds, while the soil moisture stored in the deeper deposit layers provides sufficient water for root growth, and hence for the establishment of trees.

In spring 1989, a large flush of acacia seedlings was observed in the *Tamarix-Pulicaria* zone. Because no fruiting trees were growing in the immediate vicinity of the main germination site, runoff rain water flowing down Wadi Allaqi was the likely mechanism for transporting seeds from the upstream wadi. The seeds of acacia are smooth-coated and therefore unlikely to attach to animal fur or bird feathers, although they can be transported internally by animals. Wind is another important transport agent in the immediate vicinity of the main germination site. Mechanical damage to the seed, whether caused by wind, waterborne sand particle abrasion or other factors, may be an important factor in breaking seed dormancy. Such abrasion will be partly a function of the distance traveled by the seed from the parent tree prior to its deposition on the soil surface.

This 1989 germination event provided the opportunity to undertake a field-based experimental investigation of the environmental factors likely to influence acacia germination success and seedling survival under conditions currently prevailing in Wadi Allaqi. A set of experiments provided evidence that protection from grazing by large herbivores, especially camels, appears to be a factor of critical importance in ensuring acacia seedling survival. It is noteworthy that the watering and shading of seedlings appear to have at best a neutral effect, and, at worst, a negative effect on their growth.[53] However, this observation can apply only to the natural regeneration of the trees; watering is an essential requirement for plant cultivation.

Vegetation Dynamics

Vegetation dynamics are one of the most remarkable features of the ecotone in the downstream part of Wadi Allaqi. The changes in vegetation distribution can be accounted for in terms of the relationship between the fluctuations of the lake level and the patterns of overspill. Studies conducted in this area between 1984 and 2006 enabled a direct and continuous observation of the vegetation dynamics to be made in relation to the changes of water level in the lake. When the lake was filling with water and water entered the wadi, there was only one zone of annual vegetation bordering the water's edge. The width of this zone was affected by the lake water and limited by the annual maximum and minimum fluctuations in its level. Since the lake began forming in 1964, there have been two peaks of high water level (see Figure 2.12), with the first peak of 177 meters ASL occurring in 1977–1978. The volume of the lake subsequently decreased and exposed the formerly inundated land, which led to the formation of new vegetation zones dominated by perennial plants, particularly tamarisk. With the increased length of exposure periods, and hence decreased moisture in the wadi deposits, drought-resistant perennials dominated by *Pulicaria* formed the transition zone with the desert plain.

The second peak took place in 1998–2000, when the water rose to 181 meters ASL and extended further into the wadi, covering previously luxurious ecotone vegetation and the adjacent desert land. At this time only one

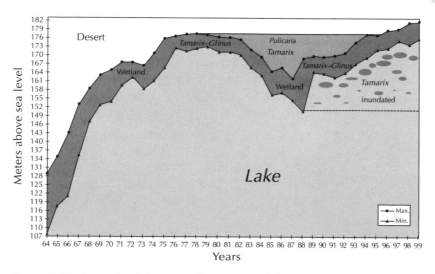

Figure 2.12: Annual minimum and maximum lake water levels over time, with area in between indicating land exposure.

vegetation zone, dominated by annuals, mainly short grasses, bordered the lakeshores. Some plants survived the inundation, particularly the water-resistant shrubs of tamarisk, the crowns of which were exposed to the air during the inundation. When the land began to be exposed again in 2001–2002, the vegetation dynamics and zone formation were enhanced by those tamarisk survivals that formed the backbone of the plant communities. Zones of annuals were soon established, followed by perennials with increasing distance from the lake. However, in the annuals' zone, *Glinus lotoides* was less abundant than in the previous land exposure, being replaced by short grasses. In the transition zone *Pulicaria crispa* was dominant, but elements of desert flora, *Cullen plicata*, *Haplophyllum tuberculatum*, and *Aerva javanica*, were greater in some locations on the desert edge.

Summary

What is abundantly clear from the discussion above is the profound impact that the impoundment of water in Lake Nasser upstream of the Aswan High Dam has had on the natural environment of Wadi Allaqi. Not only does a previously hyperarid area, with virtually no rainfall, now have a reliable supply of water all year round, but that same water has resulted in substantial modifications to the soil and vegetation resources of the area. A previously desert environment now experiences conditions somewhat similar to those found in the Nile Valley itself, with the annual fluctuations of lake water. Having set out the nature of these environmental changes in this chapter, the next chapter goes on to discuss these changes in the context of new opportunities and potentials for development.

3
New Resources, New Opportunities

Even though the vegetation of Wadi Allaqi is sparsely distributed and mainly restricted to the wadi channel, it includes plants of critical importance to the quality of life and even the survival of its inhabitants, providing food, fodder for livestock, fuel, medicine, and construction materials. Of all the recorded 127 species, 35 percent are useful in more than one context and 75 percent have some potential value.

Desert plants are valuable forage for supporting livestock production in arid lands, and Bedouin people, who live in the desert, opportunistically follow the rainfall to make use of ephemeral pasture. Animals, both domestic and wild, can graze and browse on eighty-three species growing in Wadi Allaqi. Among the most palatable of these are *Leguminosae* and grasses. The leguminous plants *Astragalus vogelii* and *Astragalus eremophilus*, *Lotononis platycarpa*, and *Cullen plicata* are the superior fodder of ephemeral pastures, while the foliage and pods of *Acacia tortilis* make an important contribution to the diet of both wild and domestic animals during prolonged dry periods. Another leguminous plant, *Crotalaria aegyptiaca*, is grazed by camels and gazelles but is poisonous to sheep and goats. Selective use of plants by animals provides an example of resource portioning among grazing animals. The camel's diet is of a wider variety than other domestic animals. They can graze on twenty-three species that are usually avoided by other animals, in addition to the highly palatable species.

Nearly half of the plant species found in Wadi Allaqi are believed to have some medicinal value. The most important medicinal plants found here are *Citrullus colocynthis*, *Senna alexandrina*, *Solenostemma arghel*, *Cleome droserifolia*, *Salvadora persica*, and *Balanites aegyptiaca*.

Almost every species of shrub, tree and woody perennial is used as fuel for cooking and heating at night during the cool winter. Among the most important sources of fuel wood are acacia and *Balanites aegyptiaca*, which grow abundantly in the upstream part of the Wadi Allaqi. Bedouin usually use only the dry branches and the dry plants for fuel, but the practices of strangers in the desert may differ; travelers, workers in the mines, and other transients cut green plants when they cannot find dry ones. Cutting live wood in the desert reduces the browsing availability and causes erosion. Reduction of woody cover is already being experienced in rural areas elsewhere in Egypt where the demand for fuel wood exceeds supply.

Acacia is a major resource for charcoal production. Charcoal yields from each tree can vary from 50 to 250 kilograms, or even more when trees are large. The use of trees for timber, however, is very limited in Wadi Allaqi. Bedouin very rarely cut trees, preferring to use them for other purposes such as browsing or charcoal. Only in the upstream part of the wadi do some Bishari people use dry wood to build houses. Examples of important timber plants are *Balanites aegyptiaca*, *Hyphaene thebaica*, and *Salvadora persica*. The wood of the last two is termite resistant and widely used in Upper Egypt for building purposes and for making tools.

Fruit, flowers, and the vegetative parts of some plants are important sources of vitamins, constituting an essential part of the diet of the Bedouin. Fruit of *Balanites aegyptiaca*, *Hyphaene thebaica*, *Ziziphus spina-christi*, *Maerua crassifolia*, and *Capparis decidua* are frequently consumed by people in Wadi Allaqi. Herbal drinks derived from *Pulicaria incisa* and *Cymbopogon proximus* are well-known all over Egypt; *Portulaca oleracea*, *Asphodelus fistulosus*, and *Launaea capitata* are used in salads. The grains of *Panicum turgidum* are collected and used for human consumption, especially in dry years when cultivated crops may fail. The soft fruit, young twigs, seeds, and tuberous root of *Leptadenia pyrotechnica* are eaten by nomads.

Desert plants are used in producing oil and fiber, tanning leather, and in making mats, baskets, and ropes. Among the economically important plants is *Medemia argun*, the leaves of which are used by Bedouin for making mats that cover the *kugra* (shelter). *Lupinus varius* ssp. *orientalis* plays an important role in nitrogen fixation, the roots of this plant being rich in nodules. In addition, trees and shrubs growing in the wadi provide shade, fix nutrients in the soil, and help to prevent soil erosion.

Plant Uses in Downstream Wadi Allaqi
With the formation of the lake, vegetation became established on the flood plain in the downstream part of Wadi Allaqi, in sharp contrast to the thinly spread plants in the main wadi channel beyond the flood plain. The lush plant growth, together with the available lake water, have attracted people

with their livestock to settle in this area and make use of the accessible plants. The vegetation in the flood plain comprises a small number of plant species that differ floristically from the wadi vegetation outside the lake's influence. About thirty plant species are successfully established in this area, but only five of these are abundant.

Tamarisk is the only species that produces a huge biomass. This is a new resource that only emerged several years after inundation and became plentiful in the early 1980s after the lake water receded. Because of its abundance and easy access, tamarisk, together with other plants growing here, play an essential role in the livelihoods of people who have settled or semi-settled in this area, providing essential grazing for domestic animals, firewood, and material for preparing the *kogra*.

There are no wild plants in the downstream part of the wadi suitable for direct human consumption except the fruit of *Balanites aegyptiaca*, which naturally and sporadically occurs in this area, as well as being cultivated by Bedouin on small plots and on the experimental farm managed by the Unit for Environmental Studies and Development of the South Valley University in Aswan. The flesh of the ripe fruit is eaten straight from the tree or dried for later consumption. The pulp is especially rich in carbohydrates (including sugars, which form up to 40 percent of the content) and vitamins (vitamin C in particular). The fruit is a particularly beneficial addition to the diet of children.[1]

Figure 3.1: Shelter built by Bedouin from tamarisk wood to provide shade and store goods. Photograph by John Briggs.

Fuel Wood[2]

Tamarisk is the shrub most frequently used for fuel wood, not only by Bedouin, but also by workers in the local mines, fishermen, and other people visiting or living in Wadi Allaqi. The downstream part of the wadi comprises dense tamarisk groves, and all households are situated in, or adjacent to, one of the groves. These provide shade as well as a source of fuel, which is easy to collect when it is both nearby and abundant.[3] Usually the women and children collect branches of tamarisk, which they tie in bundles and carry back to the household for domestic use. Most bushes from which firewood is collected have already been grazed by animals, so the branches are dry and light to carry. However, in periods when the water rises in the lake, most of the tamarisk shrubs are submerged and only their tops appear above water. This occurs most often in winter when the lake level is higher, the weather is cooler, and the demand for fuel rises. Due to inundation, the collection of firewood becomes quite a difficult job and men help in its collection. People have to go into the cold water to collect branches of tamarisk and let them dry for a few days before use; households use the wood more carefully than when it is abundant and easier to obtain during the summer months, after the annual inundation has retreated.

When they compare useful firewood sources, Bedouin prefer acacia because its biomass is more efficient than tamarisk for fuel purposes.[4] The latter has a high moisture content of 32.9 percent of its air dry weight, and a calorific value of 17.7 kilojoules per gram, determined as gross heat of combustion. In comparison, the moisture content of dry wood of *Acacia ehrenbergiana* is 29.5 percent and its calorific value 20.5 kilojoules per gram.[5] After the wood is burned, its ash is highly valued as an agricultural fertilizer. Tamarisk wood is characterized by a very high ash content of 4.6 percent of its dry weight, while the amount of ash is much less in acacia at only 2.9 percent of its dry weight. In spite of tamarisk producing a high quantity of ash, it contains only small amounts of carbon (0.3 percent) and of phosphorus (0.79 percent), and traces of nitrogen, while significant amounts of sulfur (11.4 percent) have been detected, indicating that tamarisk ash has little value as a fertilizer. In addition, its high iron content (2415 mg/kg ash) contributes to the ash's poor quality for fertilizer use. However, the carbon and phosphorus content of *Acacia ehrenbergiana* (3.45 percent and 3.89 percent respectively) is higher than that of tamarisk ash, while containing no sulfur.

Despite acacia wood's ability to produce more heat and burning longer than tamarisk, the latter is still the most heavily used domestic firewood species in the downstream area. Acacia, which is considered an excellent domestic fuel, is rarely used for this purpose because of its high commercial value for charcoal production and the presence of alternative sources of more easily obtained fuel.

Both tamarisk and acacia wood are used sparingly by local people in downstream Wadi Allaqi, who use it for cooking but not for heating, even on cold winter days (although it is used for heating in winter in the upstream areas). Usually Bedouin use an open fire for cooking, digging a small hole in the ground and surrounding it with stones. After having been used for cooking, the wood is not left to burn out, but rather the partially burned branches are put out in the sand and preserved for later use. Carefully extinguishing partially burned wood also prevents fires from spreading and setting fire to property, especially tents and livestock enclosures that are often near cooking areas, and also harming children who may play near a cooking hearth.

Figure 3.2: Acacia wood is used for making *gabana* (a special mix of coffee and spices), an important ritual in Bedouin life in the south of Egypt's Eastern Desert. Photograph by John Briggs.

Fodder Plants[6]

For Bedouin semi-settled in downstream Wadi Allaqi with their livestock, the grazing potential of flood plain vegetation is essential for survival in this area. The perennial vegetation provides permanent pasture, in contrast to the ephemeral pasture in the surrounding desert, which depends on sporadic rainfall. There is a consensus among Bedouin that when upstream pasture is available, it constitutes the best grazing resource to fatten livestock.[7] According to the temporal availability of grazing, plants of the flood plain are divided into woody perennials, other herbaceous perennials, and annuals.

Table 3.1: The nutritional composition of plant samples collected from Wadi Allaqi area, expressed as percentage of dry weight.

Species	Moist %	Ash %	Protein %	Carb. %	Fiber %	Fat %
Trees and shrubs						
Acacia raddiana (shoot and leaves)	58.44 ±9.9	9.09 ±1.3	14.20 ±2.1	6.69 ±0.9	46.49 ·±7.9	1.23 ±0.21
Acacia raddiana (fruit)	59.20 ±8.9	5.90 ±0.8	21.20 ±3.6	11.89 ±2.0	34.93 ±4.5	1.37 ±0.18
Balanites aegyptiaca (shoot)	65.78 ±8.6	9.60 ±1.6	12.90 ±1.9	9.33 ±1.7	31.08 ±5.3	0.97 ±0.13
Balanites aegyptiaca (fruit)	13.10 ±1.8	7.59 ±1.0	8.80 ±1.1	26.64 ±4.5	24.22 ±4.1	5.10 ±0.66
Tamarix nilotica (shoot and leaves)	78.29 ±13.3	23.13 ±3.2	18.4 ±3.1	6.75 ±1.0	10.07 ±1.8	1.50 ±0.21
Perennial herbs						
Cullen plicata	77.45 ±13.9	8.71 ±1.5	17.80 ±2.7	8.20 ±1.1	39.22 ±5.1	1.33 ±0.17
Citrullus colocynthis	78.74 ±15.0	22.30 ±3.4	10.70 ±1.8	8.59 ±1.2	18.88 ±3.2	1.13 ±0.19
Annuals						
Crypsis schoenoides	70.13 ±9.8	11.53 ±1.5	9.4 ±1.4	8.62 ±1.1	25.76 ±3.4	1.73 ±0.29
Fimbristylis bisumbellata	78.12 ±14.1	20.25 ±2.8	9.9 ±1.5	12.19 ±2.1	27.38 ±4.7	0.83 ±0.15
Eragrostis aegyptiaca	71.00 ±9.9	13.15 ±2.2	12.3 ±1.6	10.44 ±1.5	40.37 ±5.7	0.63 ±0.11
Euphorbia granulata	61.26 ±10.4	6.55 ±1.2	11.40 ±1.5	10.33 ±1.8	36.60 ±6.2	2.33 ±0.30
Astragalus vogelii	78.11 ±11.7	18.72 ±3.2	19.80 ±30	7.45 ±1.3	33.31 ±4.3	0.83 ±0.15

Note: Badri and Hamed 2000; Springuel, Hussein, al-Ashri, Badri, and Hamed 2003.

Perennial woody plants provide the permanent fodder supply. Among the trees and shrubs, the dominant plant producing a huge biomass is *Tamarix nilotica* (tamarisk), which is the main grazing source in the downstream part of Wadi Allaqi. However, Bedouin consider it to be a poor quality feed, claiming that after a steady diet of tamarisk mature animals become thin, their coats become spotty, milk production lessens, and they reproduce less frequently. Nevertheless in dry years, in the absence of ephemeral pasture in the upstream wadi, tamarisk remains the main food for domestic animals. Tamarisk has a better palatability in winter than in summer because the foliage's high salt content makes the animals thirsty. The subsequent high water consumption can be harmful under hot conditions. Tamarisk is at its best as a grazing source when it is young, and particularly when it is still under water, as its leaves, twigs, and shoots are less salty at this time. However, grazing on very wet foliage presents another problem in that animals fill their stomachs with water and do not eat enough solid food. The leaves and branches of tamarisk are rich in protein content, about 18 percent (Table 3.1), which indicates a relatively high nutritional value, but the high mineral content, especially chlorine, magnesium, and sodium being 2.66 percent, 1.44 percent, and 1.68 percent of the dry weight respectively (Table 3.2), most probably contributes to the low palatability of this plant.

The role in fodder production of other trees that grow in solitary stands in the downstream part of Wadi Allaqi should not be ignored, despite their small numbers. Among these are acacia trees (*Acacia tortilis* subsp. *raddiana*, *A. ehrenbergiana*, *A. nilotica*, and *Faidherbia albida*). Acacias are of considerable importance as fodder plants in the desert. They are the drought reserve fodder, which is eaten by stock at times when other feed is very scarce. Their ripe pods, called *ollaf*, are important fodder for domestic and wild animals, particularly during the dry summer months. Locals shake down the foliage and fruit with a traditional forked stick *(khataf)*, but do not cut green acacia branches. In the upstream part of the wadi, however, the Bedouin still practice the old tradition of shaping the tree crowns *(iwak)*, that is, the cutting of young branches as fodder for their goats and sheep when no other grazing plants are available.[8] The importance of acacias in the livelihood of pastoralists is reflected in the fact that several Bedouin families maintain their own small gardens of acacia trees in the downstream part of Wadi Allaqi.

Table 3.2: Mineral content of plant samples collected from Wadi Allaqi area, expressed as percentage of dry weight (source: Badri and Hamed 2000).

Species	P %	K %	Na %	Ca %	Mn %	Cl %
Trees and shrubs						
Acacia raddiana (shoot and leaves)	0.20 ±0.04	0.92 ±0.13	0.16 ±0.03	0.80 ±0.14	0.42 ±0.05	0.53 ±0.10
Acacia raddiana (Fruit)	0.15 ±0.03	0.92 ±0.16	0.13 ±0.02	1.00 ±0.13	0.48 ±0.08	0.71 ±0.10
Balanites aegyptiaca (shoot)	0.12 ±0.02	1.75 ±0.30	1.50 ±0.20	1.40 ±0.27	0.54 ±0.10	2.13 ±0.36
Balanites aegyptiaca (fruit)	0.13 ±0.02	1.41 ±0.18	0.18 ±0.03	0.10 ±0.02	0.24 ±0.04	0.71 ±0.09
Tamarix nilotica (shoot and leaves)	0.28 ±0.04	2.01 ±0.28	1.68 ±0.29	1.60 ±0.29	1.44 ±0.20	2.66 ±0.37
Perennial herbs						
Cullen plicata	0.26 ±0.03	2.50 ±0.43	0.19 ±0.03	0.60 ±0.08	0.42 ±0.05	0.53 ±0.09
Citrullus colocynthis	0.23 ±0.03	2.01 ±0.34	0.19 ±0.04	3.70 ±0.63	0.66 ±0.09	0.89 ±0.13
Annuals						
Crypsis schoenoides	0.18 ±0.03	1.31 ±0.20	0.43 ±0.06	0.28 ±0.05	0.31 ±0.05	0.89 ±0.12
Fimbristylis bisumbellata	0.18 ±0.03	1.41 ±0.24	0.27 ±0.05	0.60 ±0.10	0.54 ±0.09	1.07 ±0.15
Eragrostis aegyptiaca	0.21 ±0.03	1.12 ±0.20	0.37 ±0.07	0.70 ±0.10	0.54 ±0.10	0.36 ±0.06
Euphorbia granulata	0.20 ±0.03	1.51 ±0.29	0.21 ±0.04	0.50 ±0.09	0.42 ±0.07	0.53 ±0.10
Astragalus vogelii	0.27 ±0.05	1.81 ±0.27	0.18 ±0.03	1.00 ±0.15	0.84 ±0.14	0.53 ±0.08

Note: Plant samples were collected in May 1997 from different locations to cover all range of micro-habitats in sampled area. All collected plants samples (not less than ten samples) were mixed together and four plant samples were randomly taken from this mixture for chemical analyses.

The foliage and fruit of acacia trees are nutritious and contain high values of protein, ranging from 14 percent in young shoots and 21 percent in fruit (Table 3.1), being higher in the wet season and lower in the dry season.[9] Another useful tree for livestock farmers is *Balanites aegyptiaca*, which is cultivated on the South Valley University farm and by some Bedouin on their small plots in downstream Wadi Allaqi. It provides shade for livestock, bearing a dense cover of dark green leaves, and often keeps its leaves during the dry season when most other trees are bare. Although it is not highly regarded as a source of fodder and livestock will often feed on other plants in preference to *Balanites aegyptiaca*, animals do not hesitate to feed on the tree when other fodder sources are scarce, especially toward the end of the dry season.

The foliage of *Balanites aegyptiaca* contains less protein (about 12 percent, Table 3.1) than tamarisk and acacia, but it is rich in carbohydrates (about 9 percent, Table 3.1). The fresh, green young shoots are also palatable, but have less nutritional value than the leaves. The leaves and young shoots of *Balanites aegyptiaca* have a high mineral content, some of which (e.g., potassium and phosphorous) may be beneficial to animals while others are harmful (e.g., sodium and chlorine, Table 3.2). The most valuable part of *Balanites aegyptiaca* is the edible and nutritious fruit. The fruit is particularly rich in carbohydrates (26 percent), but poor in protein (8 percent). The seed itself (kernel), which is enclosed within the hard woody stone, is very rich in fats (48 percent) and crude protein (27 percent, Table 3.3).

Table 3.3: Nutritional composition of fruit pulp and seed kernel of *Balanites aegyptiaca*, expressed as percentage of dry weight.

Component	Pulp	Kernel	Fruits
Total carbohydrate content	88	20.8	26.64
Crude protein	1.2–6.6	27.6	8.80
Fats	0.1–0.4	48.3	5.10
Crude fiber	0.4–4.4	0.3	24.22
Vitamin C	0.9–1.6	–	not detected
Ash	2.4–6.9	3.0	7.59

Sources: Abu-Al-Futuh (1983); Booth and Wickens (1988); Giffard (1974); National Research Council (1983); Nour, Ahmed, and Abdel-Gayoum (1985).

Perennial herbs that are abundant in the downstream part of the wadi, particularly in the transition flood-desert zone, are *Pulicaria crispa* and *Aerva javanica*, both of which have low grazing value. However, according to the local Bedouin, the former can be grazed by sheep and goats, while the dry flowers of the latter are good for sheep during their lactation period. Less abundant are *Cullen plicata*, *Citrullus colocynthis*, and *Senna alexandrina*. The latter two plants are usually avoided by livestock, although the partly eaten fruit of *Citrullus* has often been observed in Wadi Allaqi and most probably has been eaten by wild animals. *Cullen plicata* merits special attention in the group of perennial fodder plants. This drought-resistant, leguminous herb with a high grazing value[10] is abundant, having a high biomass and forming an important component of the desert pasture for up to two years after a rain event. It can produce new offspring even following a light shower. Chemical analyses of *Cullen plicata* forage show it as having high nutritional values with 17 percent of protein (Table 3.1) and a considerable quantity of minerals, especially potassium (2.5 percent of dry weight, Table 3.2). In spite of their low palatability, *Senna alexandrina*, *Pulicaria crispa*, and *Aerva javanica* have a moderate protein content in the winter season (13 percent for all species), which decreases in the dry season (9 percent, 7 percent and 4 percent respectively).[11] Analyses of *Citrullus* samples taken in the dry summer period show a lower protein content (10 percent, Table 3.1) compared with other species, but a high amount of minerals, especially calcium (above 3 percent, Table 3.2).

Annuals and short-living perennial plants are seasonally available in the downstream part of Wadi Allaqi. There, during the spring-summer season, annual pasture is available on the exposed shore of Lake Nasser. Short-living annual plants, such as *Eragrostis aegyptiaca*, *Fimbristylis bisumbellata*, *Crypsis schoenoides*, and *Glinus lotoides* var. *dictamnoides,* dominate the plant communities on the lakeshores. *Eragrostis aegyptiaca*, *Fimbristylis bisumbellata*, and *Crypsis schoenoides* are the most preferred fodder plants with protein contents that vary from 9 to 12 percent and moderate carbohydrate contents of 8 to 12 percent in these plants (Table 3.1). Of these three plants, only *Eragrostis* has a high fiber content, of some 40 percent. *Glinus lotoides* is the most abundant groundcover plant, with a high biomass. However, despite its reasonably high protein content (8 to 13 percent), it has a low grazing value compared with other species growing in this area. *Hyoscyamus muticus*, which also appears on the shoreline, is rich in protein (about 17 percent in the winter period), and has a high in-vitro dry matter digestibility (DMD) of above 40 percent in the winter time.[12] Small amounts of this plant are given to young animals.

Relatively few typical desert annuals are found in the downstream part of Wadi Allaqi, but these include *Astragalus vogelii* and *Euphorbia granulata,* ephemerals with a short period of growth (three to five months) that are found on the edges of the desert and flood plain after a rain event. *Astragalus vogelii* has a high grazing value, as confirmed by its high protein contents (about 19 percent, Table 3.1), and a reasonably high amount of both phosphorous and potassium (0.27 percent and 1.81 percent respectively, Table 3.2). This plant is highly valued as fodder by the Bedouin, who claim it encourages fertility and milk production in their animals. *Euphorbia granulata* is a small prostrate plant with a high fat content (2.3 percent), but because of its small biomass it makes little contribution to animal diet.

Special attention should be paid to the aquatic plant *Najas* sp. that grows in shallow water close to the shores. This plant is frequently collected by the local people (mainly women and children) from the water and dried in the sun before being fed to animals. *Najas* has moderate nutritional content,[13] but it is used in winter because of the shortage of fodder when the shoreline pasture is submerged. It is also valued by local Bedouin because it can be stored and saved for a time when other fodder plants are scarce.

Palatability varies among plant species growing in downstream Wadi Allaqi. The most palatable are short grasses growing on the shores and desert legumes, for example *Eragrostis* and *Astragalus vogelii*, which at the same time are very nutritious. However, despite the high nutritional content of some plants, they may also have low palatability and are not eaten by animals, or eaten only when other plants are not available. Palatability can be influenced by high mineral contents in grazing plants, such as *Balanites aegyptiaca*, *Tamarix nilotica*, and *Citrullus colocynthis*, which have a mineral content of above 5 percent (Table 3.2). As mentioned earlier, Bedouin have observed that tamarisk fodder makes animals very thirsty and they drink a lot of water, adversely affecting their fattening and making them heavy, moving with difficulty.

Other plants of low palatability are known for their medicinal value (*Senna alexandrina*, *Hyoscyamus muticus,* or *Glinus lotoides*) or have a strong odor (*Pulicaria crispa*). The presence of different toxic elements in plants, such as alkaloids, phenolic compounds, and particularly tannins,[14] has a negative effect on the fodder value (intake and digestibility) of plants.[15] For example, the low palatability of *Glinus*, the most abundant annual on the lakeshore, could be attributed to the high content of saponins and phenolic glycosides.[16] *Heliotropium supinum* is unpalatable, and even toxic, because of the presence of pyrrolizidine alkaloid, but it has been observed in Wadi Allaqi that camels and goats eat these plants in the absence of other forage, despite the risks. Cutting plants or their parts and drying them in direct sunlight causes the decomposition of these toxic compounds and hence improves palatability. A common Bedouin practice is to cut annuals such as *Glinus lotoides*, *Heliotropium supinum*, and *Hyoscyamus muticus*, dry them in the direct sun, and feed them to animals when there is a shortage of other

Table 3.4: Main medicinal plants in Wadi Allaqi and their most common uses (Miller and Morris 1988, Boulos 1983).

Plants with commercial value	Colloquial name	Uses
Acacia nilotica	*sant, sont*	The most commonly used parts are the pods with seeds enclosed which are used for treating coughs.
Faidherbia albida	*kharaz, haraz*	Widely used for the treatment of coughs, pneumonia, and vomiting.
Balanites aegyptiaca	*higlig, lalub*	The fruits are best known for the treatment of non-insulin-dependent diabetes; the roots and fruits are used as a laxative and purgative in treating stomach disorders and the treatment of colds, influenza, and fever. Fruits and leaves are also used for the treatment of rheumatism and skin problems. A chemical extracted from the fruit and bark of *Balanites aegyptiaca* can kill the fresh water snails that carry the bilharzia parasite.
Citrullus colocynthis	*handal*	Best known as a purgative and against snake bites and scorpion stings.
Cleome droserifolia	*afein*	Used for infantile convulsions and as a antihelminthic and counter-irritant in chronically painful joints.
Cymbopogon proximus	*halfa barr*	Widely used in Egypt as a refreshing herbal drink as well as being used in the Egyptian pharmacological industry. It is the main component of 'Proximol' *(Halphabarol)*, an antispasmodic drug with an efficient propulsive effect. 'Proximol' also has a bronchodilator effect and is used as an antiasthmatic drug.
Pulicaria incisa	*shay gabali*	The leaves, flowers, and small branches are used to make an infusion that is a very popular drink in Upper Egypt.
Salvadora persica	*arak, miswak*	Used in folk medicine as an anti-rheumatic, analgesic, stimulant, and tonic in amenorrhea. It is also used to treat swollen spleen, sores, fever, headache, stomach pains, as well as many disorders related to the respiratory tract. The edible fruit is carminative and analgesic. Twigs are used locally as toothbrushes and some drug companies produce toothpaste from extracts of this plant.

Table 3.4 contd: Main medicinal plants in Wadi Allaqi and their most common uses (Miller and Morris 1988, Boulos 1983).

Plants with commercial value	Colloquial name	Uses
Senna alexandrina	*sennamekki*	The laxative components of the leaves and pods of this plant are anthraquinone glycosides, which give it its purgative characteristics.
Solenostemma arghel	*hargal*	Used for the treatment of coughs and as a purgative.
Plants locally used		
Aerva javanica	*ara', yara', shagarat al-naga, shagarat al-ghazal*	A paste of the roots is known to be applied to facial acne, and it is believed by Bedouins that smoke from the burning of its leaves offers protection against the 'evil eye.'
Haplophyllum tuberculatum	*urn al-ghazal*	Used to relieve pain in the stomach and chest, and for headaches and toothache.
Glinus lotoides	*hashishat al-'aqrab, ghobbeira*	An antiseptic and anthelmentic, also used for the treatment of diarrhea, bilious attacks, boils, wounds and pains, and for strengthening weak children.
Hyoscyamus muticus	*sakarran*	Contains a large number of tropane alkaloids, such as atropine and hyoscine, which are widely used medicinally as medriatics, antispasmodics, and antiasthmatics.
Cullen plicata	*marmid*	Anthelmintic, antipyretic, analgesic, anti-inflammatory, diuretic, diaphoretic, and useful in bilious infections, leprosy, and menstruation disorders. It is used in folk medicine for the treatment of skin-photosensitizing activity.
Ricinus communis	*kharua'*	The best-known use of this plant is the production of castor oil from its seeds, which is an important component in soap, printing inks, candles, and many other products. In medicine, its main use is as a purgative. In traditional medicine, the leaves are used to treat wounds and the roots to treat infection caused by guinea worm.
Tamarix nilotica	*athel, athl*	Different parts of the plant are used internally as a diuretic, depurative, and sudorific; externally, they are used as a wash for skin allergies and as a carminative. Also, the branches and leaves are combined in making a bath or lotion to treat children with measles.

fodder. Similarly, they use this technique with *Aerva javanica* and the fruit of *Citrullus colocynthis*. The latter have a very bitter taste when fresh, but can be given to livestock when dry, although only in small amounts. Adam, Al-Yahya, and Al-Farhan (2001) warn that the daily use of 25g/kg of the shade-dried fruits of *Citrullus colocynthis* for forty-two days caused slight diarrhea and had other minor ill-effects on sheep health.[17]

Medicinal Plants

Many plants growing in Wadi Allaqi are known for their medicinal value (Table 3.4), but only a few have commercial value and are collected for selling in markets and to herbal shops. Some plants are used locally to treat different disorders, while others have chemical constituents that indicate their pharmaceutical potential. Plants with a commercial value mainly grow in the upstream part of Wadi Allaqi and are well known to the Allaqi residents, while those having the potential for medicinal uses abundantly grow in both the upstream and the downstream part of the wadi, representing a potentially new resource for future exploitation.

Of five acacia species growing naturally or cultivated in the downstream part of Wadi Allaqi, *Acacia nilotica* is the best known for its medicinal value. Different parts of this tree are used in traditional medicine, particularly in Upper Egypt. The most commonly used parts are the pods with seeds enclosed, which are used for treating coughs and sold in herbal shops. *Faidherbia albida*, which is also known by its previous name *Acacia albida*, is widely used for the treatment of coughs, pneumonia, and vomiting. In the driest regions of Africa, the boiled seeds are eaten by humans during famine. The gum, bark, and pods of *Acacia tortilis* subsp. *raddiana* are all used to treat different disorders. In particular, the bark of some acacia species yields gum that is used as a demulcent.

Another plant that deserves special attention is *Balanites aegyptiaca*, one of the most widely used folk medicines, not only in Egypt, but in many countries in the Horn of Africa. The seeds, fruits, and even its flowers are sold in African food markets. Different parts of the *Balanites aegyptiaca* plant are used traditionally as medicinal cures for a wide range of complaints. These vary from place to place, as local healers develop their own skills and as traditions of medicines have been built up over many generations.

Both *Citrullus colocynthis* and *Cleome droserifolia* are used as traditional medicines by the Bedouin living in Wadi Allaqi, although the medicinal significance of the former is well-known throughout Egypt and its fruit is sold widely in herbal shops. All Wadi Allaqi households regularly drink infusions of sweetened *Cymbopogon proximus* and *Pulicaria incisa*. The downstream Wadi Allaqi households purchase the *Cymbopogon proximus* herb in Aswan, while those from the upstream part of the wadi collect it in the wild. *Pulicaria incisa* is colloquially called *shay gabali* because of its strong and pleasant smell. *Salvadora persica, Senna alexandrina,* and *Cymbopogon proximus* are all well-known medicinal plants collected by Bedouin in the desert, sold in herb shops, and used in the pharmacological industry. *Solenostemma arghel* is an important source of income for the Bishari people living in upstream Wadi Allaqi, but it has now become a vulnerable species that has a limited distribution in Egypt, being under threat because of its intensive overuse.

Although known to have important medicinal properties, some plants seem to have only a limited commercial value at best. These are listed in Table 3.4 as locally used plants: *Aerva javanica, Haplophyllum tuberculatum, Glinus lotoides, Hyoscyamus muticus, Cullen plicata, Ricinus communis,* and *Tamarix nilotica*. Medicinal uses of tamarisk, as well as *Glinus lotoides* and *Cullen plicata*, appear in literature but are not used by Bedouin living in Wadi Allaqi. As the Bedouin become more familiar with these species, which have only been growing in the area for the past two decades, perhaps they will learn more of their medicinal properties.

Soil Sustainability for Cultivation

The soils of Wadi Allaqi are primarily alluvial in origin, with some aeolian deposition. Wind-blown sand usually accumulates on the wadi edges, while the central part of the wadi contains an alluvial deposit that is the most suitable for cultivation. Silt brought by the lake water that covers the wadi floor in the flood area enriches the soil fertility of this previously desert ecosystem. In this context, it should be mentioned that the fertile soil of the Nile Valley was formed by the silt brought by river water and deposited during the flood. Nowadays, the Aswan High Dam prevents much of this silt from being carried north of the dam, and virtually all the silt accumulates in the lake's flood plain. The silt layer varies in depth and decreases with distance from the lake, in accordance with the length of periods of inundation and their frequency. This silt layer provides a favorable substrate for plant growth and also acts as a film preventing the evaporation of water from the subsoil, where it is stored after inundation. Available soil moisture and rich silt substrate support the growth of fast-growing annuals in the annually flooded areas. The dense growth of annuals, such as *Glinus* spread on the soil and the fibrous roots of grasses growing close to the water's edge anchor the silt and prevent it from being blown away by the *khamasin*, (strong spring winds occurring in Egypt and Sudan), which often occur in this open area. In spite of some dry plants being blown away by wind, annuals still produce a rich biomass that increases organic matter in the soil and hence its fertility. During prolonged inundation, the formation of an anaerobic zone in the upper part of the soil profile may have important effects on the availability of some nutrients, especially trace metals that affect crop production. Studies of trace metals in soils subjected to inundation have suggested that trace metals are not available in large amounts in these soils, and certain essential trace metals, particularly iron and zinc, may be insufficient for some plant requirements; therefore, the addition of trace metal fertilizers may be necessary for some crops.[18]

The situation differs in infrequently inundated lands, where the sparse plant cover does not protect the soil surface and the thin silt layer deposited by the lake water is soon blown away by the wind. Consequently, little organic matter is added to the soil and the rapid oxidation rate of the organic material caused by high temperatures suggests a nitrogen and phosphorus deficit that in turn negatively affects soil productivity.[19]

Therefore, to evaluate the soil suitability for crop production, a soil survey was conducted on an area of 500 feddans at the edge of the flood area, at an elevation of 178–179 meters ASL.[20] The soils are mostly sandy loam to loamy sand in texture, and the surface is covered by a thin layer rich in silt and clay, except in the middle wadi, from where the soft material is removed by runoff water or blown by wind, only to accumulate on the wadi sides. Soils are non-saline with EC less than 0.7 dS cm^{-1} in 65 percent of samples, to slightly saline with EC values, in the range from 0.8 to 3 dS cm^{-1}, in 24 percent of the samples. However, in a few locations on the wadi edge the EC indicates high salinity with measurements above 3 dS cm^{-1}. Soils are alkaline, with pH values in the range from 7.4 to 8.6. These soils contain 0.4–62.5 meq/L of calcium and 0.2–12 meq/L of manganese, appropriate amounts of these elements for crop production, but they are poor in soluble potassium (0.41–2.2 meq/L). More than 90 percent of soil samples contain chlorine in quantities less than 10 meq/L, which can be easy leached by irrigation, but in a few locations there are patches of soil with high chlorine levels of up to 64.9 meq/L. Studies have revealed that 28 percent of the area is highly suitable for farming with irrigation, 60 percent is moderately suitable, and 12 percent is marginally suitable. The best quality soils are located in the central part of the wadi, while patches of poor quality soils are on the wadi sides.[21]

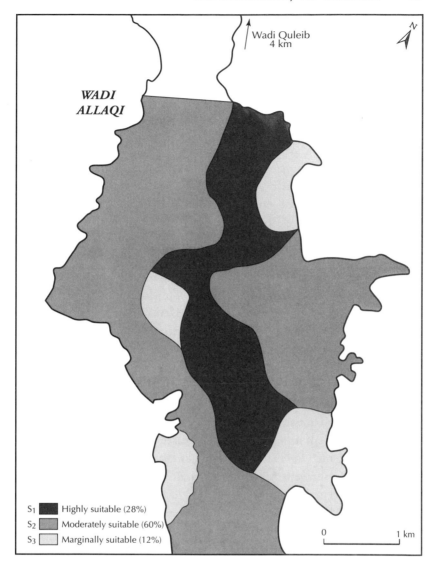

Figure 3.3: Cultivation suitability of land for sample area in Wadi Allaqi.

When considering the soil's potential productivity for cultivation, soil salinity is one of the major factors adversely affecting crop production. The soils of Wadi Allaqi outside the influence of the lake are not saline, but excess water usually brings salinity problems in hot deserts as a result of intensive water evaporation and the subsequent accumulation of salts on the soil surface. On the lakeshore in the downstream part of the wadi, chemical analyses of soil samples have revealed an accumulation of soluble salts in the surface layer (EC ranging between 0.53 to 1.81 dS cm^{-1}), particularly in the thin surface crusts, but not in the subsoil (EC range 0.12–0.24 dS cm^{-1}). This accumulation may represent the deposition of salts on the surface following the evaporation of water.[22] Away from the water's edge the soil salinity tends to decrease, although there are some 'hot spots' in the middle part of the wadi with high salinity levels in the surface layer that are associated with dense tamarisk growth. At the edge of the flood plain, where the lake's effect declines, the soils are not saline, which is reflected in the sparse growth of tamarisk. There is a risk of increasing soil salinity if tamarisk is cleared to allow cultivation, as this exposes soils with a high sodium content and, in some cases, with a high electrical conductivity.

Furthermore, the problem of salinity could be exacerbated by the irrigation-based system because of the low degree of leaching, the use of saline water from the wells, and hence salt accumulation in the irrigated soils.[23] In the Unit of Environmental Studies and Development (UESD), South Valley University experimental farm, located in Wadi Allaqi, the EC of the surface soil layer (from 0 to 10 cm in depth) increased five-fold, from 2 dS cm^{-1} to 10 dS cm^{-1}, after drip irrigation had been applied.

To use the soil resources for successful crop production, a number of management options have been proposed. First of all, plants for cultivation should be carefully selected and planted in suitable habitats. Annual plants (legume crops and other vegetables) with shallow roots will benefit from the fertile layer of silt brought by the lake water, which accumulates on the wadi floor. As shown in Chapter 2, the soil moisture stored in the soil after inundation is insufficient for plants to reach maturity (even fast-growing crops), and hence irrigation is needed.

Because the lake shoreline is not stable, water for domestic use and small-scale irrigation is obtained from shallow wells dug in the wadi bed. Water in such wells has some degree of salinity and, when crops are irrigated, salts accumulate in the topsoil where most roots are found. The deeply rooted perennials (trees and shrubs) avoid the salts accumulated in the topsoil and, once established, benefit from the subsurface water. However, there is a great risk from inundation if they are planted close to the lake, while there could be a risk of water shortage when plants are grown on high ground at a great distance from the lake. Management approaches, which are concerned with maintaining a low salt concentration in the root zone, include using appropriate irrigation techniques, weed control, ridge-furrow cropping, and covering the soil surface.[24]

Trickle irrigation is the most widely used method in the reclamation of desert areas. A small amount of water is applied to the soil surface to meet plant requirements. If the application rate matches the water utilization rate, then the upward movement of water in the soil should be small. However, no matter how good the quality of the irrigation water, it will still introduce some salts into the soil, directly into the rooting zone, where they will accumulate. Because of salinity levels found in Wadi Allaqi wells, trickle irrigation is not the most appropriate technique for crop production in this area. An alternative method, originally designed for the experimental farms in Wadi Allaqi, is subsurface irrigation. Water is supplied by underground pipes directly into the root zone of the plants. This allows better use of the water and decreases evaporation losses from the soil surface. For such a scheme to be effective, it is essential to know the rooting depth of the plants being grown. Under the very arid conditions of Wadi Allaqi, it is also unlikely that the upward movement of water and evaporation from the surface will be completely eliminated. Thus, there is still a risk of salinization of the soil in the rooting zone, which has to be monitored on a regular basis. To deal with this, a periodic flooding with low salinity water can be applied to flush out those salts that have accumulated in the surface layers of the soil and transport them below the rooting zone. This practice can be used in conjunction with trickle irrigation periodically to remove the accumulated salts. Flushing twice, at a few days' interval, can remove most of the salts from the soil.

A rigorous program of weed control helps to decrease the amount of upward movement of water in soil. Although relatively small compared with the water uptake by a crop, it may still be significant. In ridge and furrow cropping, the soil is formed into ridges and furrows prior to planting the crops, which are grown in the furrows. Water evaporates from the highest soil surface, which is at the top of the ridges, and the salts are concentrated in the soil above the plants' roots. There is still the problem of a highly saline topsoil that will have to be treated following harvesting. If the surface of the bare soil between plants is covered the amount of water lost by evaporation is lower, and hence upward movement of water and salts decreases. Any appropriate covering can be used. Plastic sheeting is often used in horticultural systems, but this is expensive and has to be brought from Aswan, and is not be suitable in Wadi Allaqi. Alternatives are plants, such as palm leaves or aquatic weeds, but not tamarisk owing to its salinity, or stones, which are plentiful in Allaqi. The main problem with this latter system of covering the soil is the likelihood of attracting scorpions and snakes.

Cultivation in Wadi Allaqi

The biological productivity of the new ecosystem (ecotone) formed in the flood plain is rich owing to the presence of lake water and high solar radiation. However, the use and management of this resource to meet the criteria of sustainable development is a very complicated process owing to high instability and inherent vulnerability of the ecotonal system. Alternating periods of flooding and exposure have a major effect on the vegetation, soil composition, and amount of water stored in the wadi fill deposits. There is no analogue for the sustainable utilization of such a fragile resource, either in Egypt or elsewhere. Therefore, any development has to be carefully planned to avoid land degradation. The development of agriculture, particularly crop production in the wadi, is the greatest cause of environmental concern. At present, three approaches to crop production occur in Wadi Allaqi. The first is shoreline cultivation for the production of commercial crops, supported by government authorities and involving private capital. The second is a research program that builds up scientific knowledge for the sustainable development of natural resources. In the context of this program, a demonstration farm has been established to promote agro-forestry industry based on the cultivation of indigenous plants. The third, but by no means least, is small-scale crop cultivation by local Bedouin mainly for domestic use, which will be discussed in greater detail in Chapter 4.

Commercial Cultivation

Water from the lake and soil enriched with silt deposits have attracted developers, especially those seeking new land for cultivation. The first attempts to reclaim the shoreline began at the end of the 1970s when the lake filled. However, large and unpredictable fluctuations of the lake's water levels, the remoteness of the area, the absence of infrastructure, and inadequate road connections to the settlements discouraged agricultural development on its shores. It was then realized that only carefully selected sites with suitable topography could be used for cultivation, while large investment should be applied to commercial crop production. Large-scale cultivation is undertaken on the western side of the lake, where shores have low slope gradients. On the eastern side, which is bounded by steep slopes, there are just patches of cultivated fields on the lakeshores. The requisite infrastructure, together with a network of irrigation canals and water pumps to obtain water from the lake, was established in the 1990s on the northern and southern sides of Wadi Allaqi at Turgoumi and Sayala-Moharraqa respectively. The areas selected have higher slope gradients than the main channel of Wadi Allaqi, consequently reducing the horizontal movement of the water when the water level of the lake rises and falls.

Cultivation strategies have been adapted to the instability of the shore. Crops are cultivated all year round, except during the two summer months (July and August) when the water level in the lake is at its lowest. Watermelon, cucumber, tomato, cantaloupe, and eggplant are the main commercial crops in the farms on the lakeshores. Watermelon and cucumber are harvested twice per year, in December and June for watermelon and in October and December for cucumber. For other crops there is only one harvesting period. Cantaloupe and eggplant are winter crops harvested in December, while tomatoes are continuously harvested from October until the end of May. Fodder plants and vegetables are cultivated for local use. Because of the specific climatic conditions on the lakeshores, with warm temperatures throughout the year, the winter cultivation of watermelons and cantaloupe have been highly successful, while in the rest of Egypt these are summer crops.

In commercial crop production, large quantities of fertilizers and pesticides are used. The sandy texture of the soil allows for a rapid leaching of the chemicals into the lake with the consequent problem of water pollution. In the absence of strict controls, residue fertilizers and pesticides are washed out into the lake with the drainage water and from the soil surface when the fields are flooded during inundation, adding to the contamination of the lake water. Uncontrolled applications of pesticide have impacted not only on aquatic biota, but also on terrestrial biota. Wild animals (rodents, foxes, and birds) have been poisoned by eating the crops and insects in the fields. Some Bedouin have also lost camels that accidentally entered the cultivated areas. During a field visit in December 2002 to Turgoumi, one of the main locations for commercial production, four camels were seen lying dead in a field, and close to them a fox family (a female with four cubs) was also dead. Partly decomposed bodies of birds were also seen in the same area. Many empty pesticide containers were thrown away in the field. The use of pesticides and chemical fertilizers on the lakeshores is forbidden by law, but without pesticides the cultivation of crops is almost impossible.[25] For this reason, farmers continuously use pesticides. Illegal uses of pesticides are periodically reported and action has been taken in each case. Other problems that face farmers are birds, both migratory and winter residents. Birds cause serious damage to crop production and illegal bird shooting also takes place.

Experiments with the Cultivation of Indigenous Plants

The cultivation of desert plants for the restoration of the habitat in the downstream part of Wadi Allaqi, and as an alternative to the exploitation of wild stocks, has been initiated for plants with potentially high economic values. The selection of plants is restricted to those deeply rooted phanerophytes (trees and shrubs) growing in the Wadi Allaqi area. After a short period in the nursery, with irrigation provided, and when the roots are well-established, the trees are expected to be self-sustaining and will naturally regenerate in the ecologically suitable habitats in which they have been planted, taking advantage of the lake water. Furthermore, the choice of the plants is based both on ecological and sociological criteria, including the plants' ability to tolerate the extreme conditions (drought and flooding) characterizing the ecotonal system. The sociological aspect relates to the value of plants in relation to their uses and is based on local/indigenous knowledge systems of Bedouin living in Wadi Allaqi.

Studies conducted by the Unit for Environmental Studies and Development team from South Valley University in Aswan have indicated that *Balanites aegyptiaca* has potentially the highest economic value of the plants growing in Wadi Allaqi. This plant grows successfully in the dry wadis of the Nubian Desert and in the Nile flood plain in Sudan. The significance of *Balanites aegyptiaca* is highlighted by the fact that it is recognized by IPALC (The International Program for Arid Land Crops) as an important desert crop that makes a high contribution to the economy of the regions where it grows. Despite its nutritional and medicinal value and other uses, *Balanites aegyptiaca* is practically unutilized in Egypt, probably because its natural growth is relatively small and scattered making it difficult to establish a *Balanites aegyptiaca* industry. An equally important tree is *Faidherbia albida*, which also has broad ecological amplitude, being tolerant of both drought and flooding.

Other trees that have been planted on the Wadi Allaqi experimental farms are *Acacia tortilis* subsp. *raddiana*, *Acacia laeta*, *Acacia nilotica*, *Tamarix aphylla*, *Ziziphus spina-christi*, *Hyphaene thebaica* (doum palm), and *Phoenix dactylifera* (date palm). All these trees grow in the Nile Valley, but their distribution is limited there by the scarcity of natural habitats. Most of these species, apart from *Acacia laeta*, *Acacia nilotica,* and the date palm *(Phoenix dactylifera)*, now grow only in ecologically suitable desert habitats. The upstream part of Wadi Allaqi that receives the runoff water from the elevated Gabal Elba region represents such a desert habitat, which is rich in woody plants. Tamarisk bushes are scattered in desert wadis where the groundwater is close to the surface. The fossil hillocks of tamarisk in Wadi Allaqi indicate its past distribution. *Acacia laeta* is a threatened tree in Egypt, but it still grows on the First Cataract Islands and in gardens of the city of Aswan. Its successful growth on the lakeshores would enrich its population in Egypt. *Acacia nilotica* was a common tree of the now submerged Nubian Nile Valley, while date palms were the main crop cultivated there. Both plants could contribute to the restoration of habitats on the lakeshores.

Experimental farms to test the growth of woody plants in a variety of habitats have been established by South Valley University in the flood plain and on the higher ground above the area inundated by the lake. The first trees were planted between 1991 and 1994 in the flood plain area near the field station to green the area and to test the growth of selected indigenous plants: *Balanites aegyptiaca, Acacia nilotica, Acacia tortilis* subsp. *raddiana, Tamarix aphylla, Faidherbia albida, Ziziphus spina-christi, doum,* and date palms. A few trees that are not indigenous to Egypt but are commonly planted in desert areas, such as *Acacia saligna, Acacia farnesiana,* and *Casuarina* sp., were also tested for sustainability in this area. At this time, the lake had reached its lowest water level (below 159 meters ASL). The cultivation plot was at a considerable distance, about seven kilometers, from the lake to escape subsequent inundation. Water was brought from the lake by water tankers and was added to the trees manually by hose once or twice a week while the plants were young (for five to six months after planting), after which irrigation was reduced to twice a month. The fastest growing trees were *Acacia saligna, Acacia farnesiana,* and *Casuarina* sp., which reached heights of two to three meters in three years. However, *Acacia saligna* did not live long and most trees died a few months after the irrigation was reduced. *Ziziphus spina-christi* grew well (up to 150 centimeters high) in the first two years, but most trees died in the third year. In two years, *Acacia nilotica* grew up to one-and-a-half meters in height, while *Acacia raddiana* remained only twenty-five centimeters high. The height of *Faidherbia albida* remained low in the first two years of its growth, but the tree rapidly increased in size in the third year, producing new individuals from suckers and becoming a strong competitor to nearby trees. *Tamarix aphylla* was the only tree planted from cuttings, while all other plants were grown from seed. Tamarisk is a fast-growing tree, reaching a height of two-and-a-half meters in two years. Despite its slow growth, *Balanites aegyptiaca* was very successful; in its first year its growth was about fifty centimeters and about seventy centimeters a year in the next two years, with five-year-old trees reaching a height of three meters. The most vigorous growth of *Balanites aegyptiaca* was near domestic sewage, suggesting that it absorbs excess water and nutrients from the sewage to sustain growth. *Doum* and date palms were among the last trees planted near the field station in 1994. Both are very slow-growing plants; in two years the highest *doum* palm was seventy centimeters, while many individuals remained only ten centimeters high. Date palm individuals varied in height from twenty-five to fifty-five centimeters two years after planting.

In 1996, with a sudden rise of lake water, the area was inundated and remained covered by water for almost six years with periodic exposure during summer time. Trees responded differently to the flood. Both *doum* and date palms died shortly after inundation, as well as *Acacia raddiana*. *Balanites aegyptiaca* and *Acacia farnesiana* survived one year, but eventually died with prolonged inundation. *Acacia nilotica*, *Tamarix aphylla*, and *Faidherbia albida* survived all six years' of inundation, and as soon as land was exposed the trees increased in size, forming well-defined trunks and crowns.

The experiment was continued, with *Balanites aegyptiaca* and acacia *(Faidherbia albida)* selected as the most promising trees for desert farming. *Balanites aegyptiaca* was planted with acacia, which, hosting nitrogen-fixing bacteria on its roots, improves soil fertility, particularly by enriching soil with nitrogen. The plot for growing trees was established in the wadi bed at 181 meters ASL elevation, in areas that were not inundated by the lake water prior to the farm's establishment in 1997. About 600 seedlings of *Balanites aegyptiaca* and 200 of *Faidherbia albida* were planted on four acres surrounded by a fence. Two types of irrigation were tested on this farm. The first irrigation system was based on the drip irrigation scheme, but to avoid the installation of filters the drips were removed and water was added to the soil surface directly from the tubes. The second was the subsurface irrigation system, which was originally designed for watering desert trees. The percolation of water downward through the sandy soil is designed to facilitate the elongation of the roots. As soon as the roots reach the underground water, irrigation ceases. PVC tubes, four-inches in diameter and open at the top and the bottom, were vertically installed in the ground to a depth of one meter, while half a meter of the tube remained above the surface (Figure 3.4). Three-to four-month-old seedlings of *Balanites aegyptiaca* and *Faidherbia* were planted close to the tubes. At the beginning of growth the surface irrigation was applied to allow the seedlings to establish themselves in the soil and to enlarge the root system. When new foliage appeared, which took about two to three months, water was added directly to the tube, bringing water close to the roots of the trees. The tubes were filled to the top twice a week. With a tube diameter of ten centimeters and length of 150 centimeters, each tree received about twelve liters of water twice a week, totaling about ninety-six

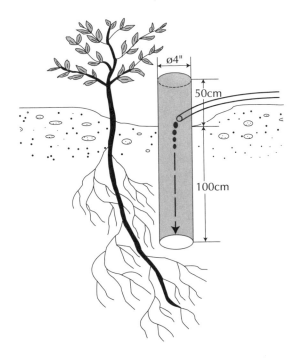

Figure 3.4: Experimental subsurface irrigation method.

liters monthly. Two hundred trees planted in one acre require nineteen cubic meters of water monthly, which is a very small amount compared with the water requirements of the main crops in Egypt.[26]

The survival rate of fifteen-month-old seedlings was high, above 95 percent. One and a half years after planting, the young trees reached a height of above one meter, with an average annual growth of above sixty centimeters. However, some individuals reached a height of above two meters; maximum heights of 2.42 meters and 3.50 meters were recorded for *Balanites aegyptiaca* and *Faidherbia albida* respectively. The fastest growth of the trees was observed in that part of the farm where sub-surface irrigation was applied. The mean height of trees under subsurface irrigation was 1.51 meters, while trees were smaller with surface irrigation, the mean height being only 1.36 meters. Subsurface irrigation supports a better growth of trees than does drip irrigation, as evidenced by the root behavior rather than the shoot growth. The first effort of a young desert tree is to grow a deep root system that is able to make the best use of the available moisture in the soil. The growth of branches above the ground is slow while the root system is being established, and the tree can remain in dwarf form for many years before reaching maturity.

The architecture of the roots of *Balanites aegyptiaca* under the surface and subsurface irrigation varies greatly. Under the subsurface irrigation, the main tap root penetrates deeply downward and begins to produce lateral roots to a depth of one meter where water is added through the irrigation tube. The volume of the soil occupied by the roots is about 1.5 meters in the vertical dimension, and less than half a meter in the horizontal dimension.

Under conditions of surface irrigation, the tap root is short and gives rise to many lateral branches that spread horizontally at a depth of two to fifty centimeters. Roots are often rounded in close proximity to the place where water is applied. The volume of the soil occupied by roots mainly extends horizontally (1.5–2 meters), while vertically it extends up to fifty centimeters. By applying subsurface tube irrigation, the vertical dimension of root distribution is greater than its horizontal distribution. This has important implications. First of all, deeply rooted trees can absorb the subsurface water and hence, when mature, will need little or even no irrigation. Second, the roots with a greater relative vertical distribution occupy a smaller area compared with horizontally spread roots; hence more trees can be planted on the same area. This will favorably affect the cost of a farm in places where protection from grazing is needed.

The South Valley University experimental farm was also inundated in the autumn of 1999, when water in the lake rose and reached its highest recorded peak ever of above 181 ASL. As a result of the lake's annual water fluctuation, the farm was submerged during autumn and winter, and exposed in spring and summer, for the following four years. *Balanites aegyptiaca* survived the first year of inundation but died in the second year. *Faidherbia albida* behaved differently, with 90 percent survival after two periods of flooding. Not only trees with their crowns above the water, but even small individuals, below fifty centimeters in height, and completely covered by water, endured the long inundation and retained green leaves. In 2002, when water in the lake retreated and the farm was exposed, the survival rate of acacia was 70 percent. Most of the acacia trees were less than one meter high, twelve trees were above three meters, and only one fruiting tree was five meters in height. Where land had been exposed, acacia grew very quickly by absorbing water that remained in the soil after inundation. In 2006, a few trees in the most favorable locations reached a height of eight to nine meters, many reaching five to six meters, while about fifty of the total 110 surviving trees remained below one meter. Vegetative reproduction, by

Figure 3.5: Young Ababda men who work on the South Valley University experimental farm in Wadi Allaqi irrigating *Balanites aegyptiaca* trees. Photograph by Irina Springuel.

producing new individuals from suckers, favored the regeneration of the acacia population on this farm.

Another experimental plot was established on the high ground of 186 meters ASL to evaluate the prospect of restoring desert habitats and improving the productivity of desert vegetation outside the flood area. Wadi Umm Ashira, the downstream tributary of Wadi Allaqi, which is close to the lake but with a high gradient and hence outside the area of flooding, was selected for this farm. Work began with the planting of 210 seedlings of *Balanites aegyptiaca* in the spring of 1999. In the following two years the farm was extended by planting more seedlings of *Balanites aegyptiaca* and *Faidherbia albida*, together with seedlings of acacia (*Acacia laeta*), henna (*Lawsonia inermis*), *Ziziphus spina-christi*, tamarisk, *doum* and date palms. Surface irrigation was used and water was brought by water tankers from the lake. Small plants were watered twice a week. However, during the hottest period watering was increased to three or even four times a week. From 2003, irrigation was decreased to once a week in summer, while in winter water was added only every two weeks.

In one year *Tamarix aphylla* died, *Ziziphus spina-christi*, *doum*, and date palms survived for two years, and only two individuals of *Acacia laeta* survived for five years. Of all the plants cultivated, only *Balanites aegyptiaca*, *Faidherbia albida*, and henna were successful. Five years after planting, the survival of *Balanites* was more than 60 percent, *Faidherbia albida* about 40 percent, and henna 30 percent. Only a few individuals of *Balanites aegyptiaca* reached the height of four to five meters in five years and produced fruit, while most trees remained below two meters high. All *Faidherbia albida* trees remained dwarf in the range of fifty to 150 centimeters high, while henna grew well, reaching a height of one or two meters in three years and flowering in five years.

All the experiments on the Wadi Allaqi farms point toward *Faidherbia albida* as the most appropriate tree for improving the productivity of natural vegetation and as the most appropriate for agro-forestry development in the flood area. This tree can withstand long inundation, while also being a drought-resistant plant that grows in desert wadis. *Tamarix aphylla* and *Acacia nilotica* are other trees growing well in flooded areas. The latter is of particular importance for Bedouin, providing a major fodder source for their animals. A few households have gardens in which they have cultivated acacia trees. However, it should be taken into consideration that *Acacia nilotica* is a typical riverine tree growing in the Nile Valley and cannot survive water shortage.

Another suitable tree for agro-forestry is *Balanites aegyptiaca*, whose best habitats are on the edge of the desert and flood plain. While growing quickly in a flooded area it cannot withstand prolonged inundation, and hence it is not recommended for development in the flood plain. The growth of *Balanites aegyptiaca* in the desert is slow but, being deeply rooted, this tree can absorb seepage water from the lake to sustain its growth without irrigation. Once established it can tolerate drought and becomes an important component of the desert ecosystem by supporting local Bedouin, providing fodder for their animals and for wildlife. *Balanites aegyptiaca* trees begin to fruit at about five to seven years of age and reach maturity in twenty-five years; they can be long-lived, sometimes surviving for over a hundred years. In Sudan, where *Balanites aegyptiaca* is abundant, its fruit is used in small-scale industry for producing oil from the kernel, which consists of between 30 to 60 percent of oil.

Summary

It is clear that the impacts of Lake Nasser water have not only changed the ecology of Wadi Allaqi but, more to the point, have changed that ecology in ways that are of potential benefit to the Bedouin living in the area. The resultant biomass now provides a range of exploitable resources. Grazing resources consequent upon lake retreat are now available from October to November until the following August, and are being increasingly supplemented by the cultivation of small amounts of fodder crops, irrigated by either lake water directly or seepage water from the lake that fills wells, sometimes at considerable distances from the lakeshore. The biomass, especially tamarisk bushes and trees, also provides firewood and building materials, as well as medicinal plants and herbs. If the soils of Wadi Allaqi are used with care and are maintained properly, they are more than capable of supporting cultivation on small farms. Against this background of opportunity, the next chapter explains how the Bedouin of Wadi Allaqi have responded to these new resource opportunities by modifying and developing their livelihood systems.

4

Sustainable Livelihoods and Natural Resource Management: Challenges for the Bedouin[1]

The Bedouin in Wadi Allaqi have developed a livelihood system comprising five key elements (Figure 4.1). Crucially, each of the elements makes use of the available resource opportunities, but does so in a managed and sustainable manner. This is, of course, vital as the Bedouin of Wadi Allaqi live in a highly fragile and marginal physical environment. Mismanagement would only result in disaster, and the Bedouin are only too aware of this. There is significantly a marked seasonal dimension to how Bedouin manage and use the resources, summarized in Figure 4.1.

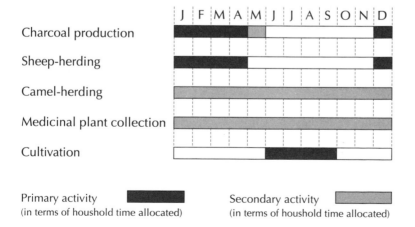

Figure 4.1: Agricultural calendar.

Livestock transhumance, particularly of sheep, in combination with charcoal making dominates the winter part of the calendar. This is associated with the availability of ephemeral grazing in the hills to the south and east of Wadi Allaqi. In these areas there are also relatively abundant resources of acacia trees, which are the Bedouin's preferred species for charcoal production. While Bedouin are away with the sheep in the hills, they take the opportunity to make charcoal at the same time. The two economic activities are, therefore, very closely related to each other as far as Wadi Allaqi Bedouin are concerned.

Cultivation, on the other hand, is predominantly a summer activity, although since the late 1990s it has extended throughout the year among some households. This reflects the fact that during the summer most Allaqi Bedouin are back living in the wadi itself because the desert beyond is too hot and uncomfortable for long periods in comparison. Labor, therefore, is available. The timing of planting is crucial because Bedouin have to make predictions about lake water retreats and advances. If Bedouin get things wrong, this can result in plots being inundated too early and the crops lost; or in the lake water being too distant, resulting in groundwater being too deep to be exploited by the wells near the plots. It is a very difficult balancing act. Camel herding and medicinal plant collection, although important economic activities for most if not all households, are secondary activities in terms of the household labor allocated to them. Perhaps because of this, they are activities that take place throughout the year.

This calendar is influenced very much by environmental variables, the first of which includes monthly variations in the levels of Lake Nasser. In a typical year, there is a difference of some six to seven meters in lake water depth between the lowest and highest lake levels, usually in the months of July and November respectively. In Wadi Allaqi, because of the relatively shallow gradient on the wadi floor, a vertical difference of six to seven meters in lake water depth can result in a horizontal difference in inundated land of ten to fifteen kilometers. Clearly, as the flood recedes from November onwards, the timing of any crop planting becomes critical. The second environmental variable is that of the incidence of winter rainfall in the Red Sea Hills to the south and east of Wadi Allaqi. In these areas, winter rains are common from mid-December until early February. Even though the annual rains in these areas average only fifty millimeters, this is enough to stimulate the growth of ephemeral grasses, herbs and small shrubs for grazing in the hills, as well as to provide sub-surface, and sometimes surface, runoff downstream in Wadi Allaqi itself. Significantly, however, there have been successive years of below average rainfall since the mid-1990s, and in some years no appreciable rainfall at all. The result has been that there has been far less transhumance of Allaqi Bedouin to these areas with their flocks. The third factor is that of ambient temperatures, particularly in the summer months when air temperatures can become oppressively high, with mean daily highs of 45° C, and temperatures exceeding 50° C on many days. During the heat of the summer, Bedouin tend to locate their settlements on the wadi floor and relatively near to the lakeshore to take advantage of the (relatively) cooling winds off Lake Nasser. These environmental rhythms, therefore, clearly influence the nature of the agricultural calendar.

Small Livestock Production

Small livestock production of sheep and goats is the major economic activity of Wadi Allaqi Bedouin, and uses, along with cultivation, the largest amount of labor resources. Up until the late 1990s, the location of sheep production was divided between Wadi Allaqi and the Red Sea Hills to the south and east. Typically, sheep were taken during the winter months for periods of two to three months at a time to graze in the hill areas, sometimes over distances of up to 150 kilometers. This had a number of advantages. First, sheepherders were able to use the opportunistic and ephemeral grazing available in the hills, even after the briefest of showers. Information on the incidence of rainfall, and hence the subsequent location of grazing a few days afterward, reached Wadi Allaqi in a number of ways, including from traveling charcoal producers and sellers, camel drovers, returning sheep grazers, and even relatives who still lived in the hills. In addition, observed physical attributes also give Allaqi residents clues as to the availability and locations of hill grazing. Types of cloud formations seen in the distance, surface water flows in wadis, and even sub-surface flows that stimulate a greening of the visible vegetation all provide information on hill grazing, even at considerable distances from Wadi Allaqi itself. Whereas there are well-established, sometimes complex, and generally well-observed ownership rules associated with permanent vegetation resources, primarily trees, there are no such restrictions with ephemeral grazing. Access to such grazing is on a first-come first-served basis, and Bedouin adopt the rather pragmatic view that when such grazing becomes available it should not be wasted, even though they themselves may be too late to access it themselves.

Second, moving the sheep to the hill areas provides an opportunity for the grazing areas in Wadi Allaqi itself to recover. This was especially marked in the 1980s and 1990s, when it was common for most sheep in the wadi to be moved except for those that were heavily pregnant, had just lambed, or were too ill or too young to be walked the distances to the hill grazing. With the onset of the drought in the late 1990s, however, this element of the production system has steadily reduced in importance, such that by 2005 it no longer formed part of the production cycle. However, this was not just the result of drought and reduced grazing in the hills. At the same time other social and economic changes were taking place among Allaqi Bedouin.

Settlement by Bedouin along the lakeshore in Wadi Allaqi is a relatively recent phenomenon, most of them arriving only from the mid-1970s onwards. For them, the lake provided a new resource opportunity, but they still held both an economic and emotional attachment to the desert (the *gibal*, literally 'the hills'; sing. *gabal*) 'out there.' Although they were increasingly settled in Wadi Allaqi, Bedouin men in particular still felt a need to be attached to the "desert proper," as one of them called it. Taking sheep to the hills in the winter provided that opportunity. However, as Bedouin became more settled over the years, became more accustomed to the advantages of a reliable, year-round supply of water from Lake Nasser, and developed greater understandings of the Wadi Allaqi resource base, so the attraction of the desert, however 'proper,' started to wane. This was reinforced by the added attractions of Aswan being only 180 kilometers away, and, after the completion of the asphalt road from Aswan to within twenty kilometers of Wadi Allaqi in the early 1990s, being very accessible as a market for the sale of sheep and other products. In addition, Bedouin women became increasingly resistant about returning to the desert, as their lives were much more attractive in Wadi Allaqi, with guaranteed water and access to rudimentary health care for their children. Perhaps an increasingly important key element in this is the fact that there is now growing up in Wadi Allaqi a generation that has little or no knowledge of the desert beyond it and, in some cases, has even less desire to spend time there. The drought, therefore, has been an important element in the collapse of the winter grazing element of the sheep production system, but it must be seen in the context of wider social and economic change.

Within Wadi Allaqi there is a considerable variation in the number of sheep owned by households. Some have as few as ten ewes, while others have as many as 350. Sadly, the number of sheep owned by a household is a key indicator of wealth differentiation. Bigger breeding flocks provide both greater opportunities for economies of scale in production and for larger numbers of lambs for sale, thus increasing income levels and hence opportunities for capital accumulation.

The availability of household labor is an important factor in influencing the number of sheep owned by a particular household, particularly as the main carers of sheep come from within the household itself. Surveys of households in Wadi Allaqi consistently show that the households themselves are responsible for looking after their own sheep. In particular, it tends to be young males who are given this task most often. Sons are introduced to looking after sheep when they are about eight or nine years old by being given responsibility for perhaps ten to fifteen sheep. As the son gets older, the number of sheep for which he is responsible is increased. Those households that have two or three sons have a distinct advantage in terms of sheep production over those households that have fewer, as this provides such households with the labor capacity to divide the flock when needed. This was certainly the case in the winter in the past, when some parts of the flock were taken to the Red Sea Hills for grazing, while other sheep were kept in Wadi Allaqi. Since the late 1990s, though, with the collapse of *gabal* grazing, there has been a narrowing of this differentiation, although this has been the poorer households getting wealthier, rather than wealthier households becoming poorer.

Although wives are also important carers of sheep, they appear not to have as big a role as husbands and sons in managing the flocks. Nonetheless, women do have an important role, especially in looking after sick sheep or those that have recently lambed and need particular care. Virtually all the care by women is undertaken in Wadi Allaqi, rarely elsewhere (see Chapter 5 for a fuller discussion). Other sources of labor are much less important, although four households do make use of paid shepherds from time to time. Unsurprisingly, these were wealthier households and owned relatively large numbers of sheep. Therefore the hiring of shepherds was not a strategy used by households with limited internal labor resources to meet this labor shortfall, but rather was a strategy by the wealthier households to release family labor for other productive economic activities.

The main lambing time in Allaqi is during the winter months, particularly the months of January, February, and March. The problem with this time of year is that the health of the pregnant ewes may not be very good, the reason being that during November and December the lake level is at its highest and so there is only a limited amount of pasture available in Allaqi itself. It may also be too early for any appreciable amount of rain in upstream Allaqi to bring on pasture growth there. Hence, the limited amount of grazing has a detrimental effect on the health of the pregnant ewes and may contribute to lower lambing percentages. Such data are notoriously difficult to collect in Bedouin economies, and hence the following can only be based on estimates. Nonetheless, surveys suggest that the lambing percentage across all households is 100.1 percent, but this masks a range from as low as 30 percent for one household to nearly 150 for another.[2] There is a marked difference, however, between those households with larger breeding flocks and those with smaller ones. For those with below-average size breeding flocks the lambing percentage is 86, whereas for those with above-average breeding flocks the equivalent figure is 104. There would appear to be clear advantages to be had from economies of scale in sheep production in Allaqi. A contributory factor may be that the owners of the larger flocks have the greater financial resources to buy in feed, if necessary, as well as having the requisite amount of household labor to take sheep for better winter grazing in the Red Sea Hills, at least in the past, although this practice is now much less frequent because of successive years of drought conditions in the hills. At this time it was worthwhile taking flocks of a hundred sheep upstream, whereas it was hardly so for only twenty or thirty sheep, given the amount of labor that had to be committed to such activities. Once again, it tended to be the poorer households that were unable to take advantage.

The most important feed resources for sheep tend to be desert vegetation species rather than those more obviously associated with the lakeshore lands. Indeed, of the seven most important feed sources, all of them scoring more than 92 percent, six are desert plants (Table 4.1). The only exception out of the seven is grass, which is associated with the emerging lakeshore areas as the lake water retreats from November/December through until the following summer. Although these desert plants are found in downstream Wadi Allaqi, they are much more common and abundant in the upstream areas. All six are either short-life perennials (*Cullen plicata*, and *Cleome chrysantha*), annual herbs (*Astragalus vogelii, Convolvulus deserti*, and *Euphorbia granulata*), or a tolerant shrub *(Pulicaria crispa)*. They are all highly responsive to moisture and hence flourish in the winter in the upstream areas, benefiting from rain of up to 150 millimeters at that time of year in these areas. These feed resources are, therefore, associated by most Bedouin with winter grazing, particularly in Haimur, Eigat, al-Fogani, Abu Fas and Ungat, all of which are located in the *gabal* to the south and southeast of downstream Wadi Allaqi. Bedouin also collect these grazing materials and transport them back to Allaqi as feed for those sheep that are not taken to the hills. The fact that they are rated so highly by most households demonstrates the importance, at least in the past, of these grazing areas for sheep in the winter. It also emphasizes, yet again, how much poorer households with smaller flocks and reduced availability of labor miss out on this resource in the production process.

Table 4.1: Importance of different feed resources for sheep.

	Very Important	Important	So-So	Not Important	Irrelevant	Score	%
Cullen plicata	18	2	0	0	0	78	97.5
Grasses	18	1	1	0	0	77	96.3
Astragalus vogelii	17	3	0	0	0	77	96.3
Convolvulus deserti	16	4	0	0	0	76	95.0
Euphorbia granulata	17	2	1	0	0	76	95.0
Pulicaria crispa	17	2	1	0	0	76	95.0
Cleome chrysantha	17	1	1	1	0	74	92.5
Glinus lotoides	12	4	4	0	0	68	85.0
Tamarix nilotica	3	6	11	0	0	52	65.0
Acacia	1	2	17	0	0	44	55.0
Maize from Aswan	1	0	3	15	1	25	31.3
Sorghum from Aswan	1	0	3	14	2	24	30.0
Balanites	0	0	5	13	2	23	28.8

Note: The data in this table are derived from asking twenty Bedouin households to score the importance of each vegetation type on a five-point scale from 'Very Important' through to 'Irrelevant.' For each 'Very Important' response, a score of four is attributed; for a response of 'Important' a score of three is attributed, and so on, with an answer of 'Irrelevant' receiving no score. The total score for each cell is then calculated, and the cells for a particular vegetation type are added up to provide an overall score. This is then expressed in the final column as a percentage of the maximum possible score, assuming that all twenty households rated the vegetation type as 'Very Important'—that is, the maximum achievable score for any vegetation type in this table is eighty. The second main source of feed is around the lakeshore. In particular, grasses are rated very highly. These grow as soon as the lakeshore starts its annual retreat. Typically, the grass will be left for about two to three weeks after first appearing, at which point sheep will be moved there for grazing. As the lake retreats, more grass becomes available and this is grazed in turn. However, as the temperatures rise in the spring it is common for new grass to be burned off by the sun before it has chance to mature for grazing. The Arabic word *nagila* is used as a collective noun for these grasses, and they include four main species, these being *Fimbrystlis bisumbellata, Eragrostis aegyptiaca, Cyperus pygmaeus,* and *Crypsis schoenoides.* Similarly, *Glinus lotoides* (*toroba* in Arabic) also grows after water retreat and offers another important source of seasonal grazing for sheep around the lakeshore. This is a ground cover plant and can grow to a height of twenty to thirty centimeters. Like grass, it is highly nutritious for sheep in its early growth stages but becomes toxic to them as it matures, and hence is then avoided. This produces the rather incongruous sight of seemingly large amounts of untouched grass growing freely around parts of the lakeshore.

More permanent sources of feed around the lake are provided by tamarisk (*abl* or *tarfa* in Arabic), acacia, and, to a much lesser extent, *Balanites aegyptiaca*. Indeed, *Balanites aegyptiaca* is seen to be the least important source of feed for sheep among Allaqi Bedouin (Table 4.1).

Tamarisk leaf, although salty, is used as a source of feed. It tends to be the new leaf growth that is selected as this is less salty than older growth. The Bedouin are well aware of this, as well as the fact that tamarisk is not particularly nutritious. If over-dependent on tamarisk for feed, milk yields from sheep fall and the incidence of intestinal diseases increases. Since 1998, however, tamarisk has been available much less as a source of sheep feed in Wadi Allaqi because of the record high inundations of Lake Nasser water. In both 1998 and 1999, the lake level exceeded the expected 180-meter level and hence inundated the entire stock of tamarisk in the wadi. This has created considerable hardship for many of the sheep-owners, in that a reliable and year-round source of feed has been eliminated. Acacia leaf is also used and scores relatively highly. Although unaffected by lake inundation, acacia is not particularly abundant in downstream Wadi Allaqi and hence has been unable to substitute in volume for the loss in tamarisk.

A final source of feed is made up of purchases of maize and/or sorghum in Aswan. As Table 4.1 shows, however, this is not an important source of feed for most Bedouin; indeed, only one respondent saw this as a very important source. Most Bedouin consider it to be an unimportant but, significantly, not irrelevant source. In other words, for many, maize and sorghum bought in Aswan is seen as a measure of last resort when there are few or no other sources of feed available. The price of grains is about LE1.5 per kilogram, although there are slight seasonal price variations around this. Unsurprisingly, it tends to be the wealthier sheep-owners who use this source the most, yet another factor reinforcing differentiation in the community.

Interestingly, the data presented in Table 4.1 were collected in the late 1990s, and since that time there has been added aquatic vegetation, and particularly *Najas* (*shilbeika* in Arabic). During the 1990s Bedouin largely ignored the lake as a resource base, except as a reliable water supply. However, successful experimentation has taken place using this aquatic vegetation species, and hence it has now become an added element in the diets of sheep and goats. It is not without its problems, however. *Shilbeika* has to be dried, and therefore loses considerable mass before being consumed by livestock. In its natural state, it contains a large amount of water, such that if it is consumed in this state, it can cause in major intestinal problems for the livestock, and subsequent loss of condition. In extreme cases, it can result in death.

Sheep provide a range of essential products for Allaqi Bedouin. Every household uses milk produced by their sheep and goats; this is clearly the most important reason for keeping them, followed closely by wool production. Interestingly, none of this milk is turned into cheese, although butter is produced, being mainly used as a type of skin moisturizer. Products that follow from slaughter, meat and skins, also have a high level of importance, but are usually extracted after the sheep are sold.

Table 4.2: Importance of sheep production problems.

	Very Important	Important	So-So	Not Important	Irrelevant	Score	%
Not enough grazing	3	2	6	1	8	31	38.8
Sheep diseases	0	0	2	18	0	22	27.5
Not enough water	0	1	7	2	10	19	23.8
Not enough feed	0	0	5	3	12	13	16.3
Animal attacks	0	0	0	13	7	13	16.3
Transport to Aswan	0	0	0	6	14	6	7.5
Unreliable market	0	0	0	1	19	1	1.3

Note: the Score column is calculated by multiplying all the responses in each line of 'Very Important' by four; all the responses of 'Important' by three; all the responses of 'So-So' by two; all the responses of 'Not Important' by one; and no score at all for a response of 'Irrelevant.' These are then summed in the Score column and then expressed as a percentage of the maximum score possible in the right-hand column of the table.

Despite the difficulties posed by a harsh physical environment in Wadi Allaqi, production problems for animals are not seen as a major issue by most Bedouin (Table 4.2). Even the lack of grazing receives a score of less than 39 percent, perhaps surprising in the light of the fact that this particular survey was carried out at the time of annual lake inundation when the availability of grazing in Wadi Allaqi is much reduced as it disappears under water. Respondents took a broader view of availability, including upstream resources, and hence only four took the view that lack of grazing was a very important problem. There is, without doubt, a clear seasonal dimension to this question. In the winter, if a sheep-owner has the labor available, grazing is not particularly problematic in the upstream areas. Similarly, in the spring, as the lake recedes, new grazing becomes available as the land around the lakeshore reemerges. It is in the late summer and autumn when grazing resources become stretched. This was especially so in 1998 and 1999 as one of the key grazing resources during this time of year, tamarisk, became almost totally inundated as a result of the record high lake levels. However, by 2004 the problem of lack of grazing had grown significantly in importance as a result of drought in the *gabal*, where grazing was virtually eliminated, and by lake inundation that covered most of the tamarisk and other bush and shrub stock. Only seasonal grasses and *toroba* are now really available, and neither has the biomass capacity, even when combined with the aquatic *shilbeika*, to provide the volume of good-quality grazing needed for secure sheep production.

Sheep diseases are also not regarded as a big issue for most of the respondents, with nobody seeing these as either important or very important. Nonetheless, the main disorders seem to be intestinal disorders and liver fluke, both of which can have a significant impact on sheep quality. Some attributed intestinal problems to the use of dried feed collected in the upstream areas or elsewhere in Allaqi itself, while others blamed the use of aquatic plants. Other problems are relatively unimportant, including both transport to Aswan and market difficulties.

Most respondents sell lambs when they are aged between one and two years and weigh between twenty-five and thirty-five kilograms, although some wait until they are slightly older, sometimes up to three years old. Typically, all the male lambs are sold for slaughter, while females are used to replace ageing ewes, or to increase the overall size of the breeding flock. The timing of sales varies throughout the year. Factors such as high market demand for sheep at the Eid al-Adha are important, as are specific household cash needs at particular times of the year for households. Lamb sales can, therefore, provide a steady supply of income for households throughout the year if, of course, they have a sufficiently large breeding flock in the first place. Cast ewes are usually sold between the ages of six and ten years, depending on the condition of their teeth. The quality of some of the grazing is such that teeth tend to be worn down quite quickly, such that by the age of six years some ewes have to be sold off for slaughter. Only a few are able to retain teeth, which allow them to graze successfully, up to the age of ten years. Most Bedouin sell their sheep in Aswan, although one or two still retain a loyalty to traders in Shalatin on the Red Sea coast.

In summary, there are three key features related to sheep production in Wadi Allaqi. First, the availability of household labor is a vitally important production factor. It is clear that the overwhelming bulk of labor committed to sheep production comes from the household itself; indeed, it is the amount of available labor that influences the extent to which a household can participate in taking sheep to winter grazing. Larger households have the labor to do this, and smaller households do not. Without doubt, access to winter grazing at some distances from downstream Allaqi improves the overall health of the breeding flock and therefore improves lambing percentages. Hence, those households with more labor available are able to increase the size of their flocks more easily, which, in turn, gives them more income from the sale of male lambs and allows them to increase further the size of the base breeding flock from retained female lambs. As a result the bigger, more economically successful households become even better off, while the poorer households generally remain significantly worse off. Sheep production, therefore, is an important wealth differentiator between households in Wadi Allaqi.

Second, the sheep production system depends very heavily on access to a range of different grazing areas. By taking sheep to winter grazing in the hills, an opportunity is created for the recovery of grazing resources in Wadi Allaqi. Hence, the present transhumance system is a sound resource management arrangement. However, yet again it is the wealthier households that are best able to benefit from it for the reasons suggested above. Although there is no private ownership of land and grazing in Wadi Allaqi, there are nonetheless agreed rights of grazing access between the various family groups resident there. The poorer households, therefore, have no option but to graze their sheep on the same areas in Wadi Allaqi, but without the benefits of the rest period enjoyed by wealthier households. The threats from overgrazing are first felt by the poorer households, therefore further reinforcing their lower economic status.

Finally, it is clear that as access to Aswan has improved in recent years, with the completion of the asphalt road from the northern ends of Wadi Umm Ashira and Wadi Quleib, the opportunities for Bedouin sheep producers to become more commercial have increased. Bedouin have produced sheep for sale for many years, but difficulties of accessibility resulted in a loss of condition of sheep on the way to market or exploitation by some sheep traders from Aswan. The asphalt road has changed this. Without doubt, this is a great opportunity for Bedouin sheep producers, but it also creates new challenges, and, in particular, the whole question of whether Allaqi sheep meet the necessary quality standards. Although there may be market advantages in selling genuine Bedouin sheep (for example, they are reputed to have a far fresher taste than Nile valley sheep), there are real disadvantages in quality as far as the consumer is concerned. This, therefore, is an area for concern with regard to the future of Bedouin sheep production in Wadi Allaqi.

Camel Herding

Camel herding constitutes an important activity, although in terms of labor allocated it is a relatively minor one. There is also considerable variation in the numbers of camels owned by households. Within the Wadi Allaqi community, this varies from between about 1,000 camels for one of the households, down to only two for the household with the fewest camels. Leaving aside the 1,000-camel household, the typical household has ten to fifteen camels at the most. Camels not only make up an economic investment in their own right, but they also provide the means for further economic activity. They are, therefore, one of the key factors in wealth differentiation between households in Wadi Allaqi. For those households with significant numbers of camels, these represent a stock of savings that can be drawn upon as capital when needed. In addition, they provide the means for further capital accumulation, especially as a key mode of transport that allows households to participate in grazing in winter grazing and charcoal production, as well as transport for selling products in Aswan and elsewhere beyond Wadi Allaqi. Those households with fewer than six or seven camels cannot afford to commit such scarce resources to these types of activities because of the other economic demands exerted on the household.

Typically, most households keep only a few of their camels in Wadi Allaqi close to the encampments, these being camels that are used for transport, camels that are sick, or camels that have recently foaled. The rest of the camels are allowed to roam freely in the desert for extended periods of time, sometimes of up to two to three years. Some Bedouin talk of camels being away for up to six years, but this figure is unsubstantiated. This roaming is quite important in maintaining the grazing resources of Wadi Allaqi itself, because if all the camels were kept in this area, there would be extreme pressure placed on these resources. The camels bear the owner's unique brand-mark (*washam* in Arabic), and even foals with their mother are likely to be branded with the same mark, even by Bedouin who may not be the owner. This system of trust is absolutely central to Bedouin, and, without it, Bedouin society, culture, and ultimately economy breaks down. Although there are increasing pressures, this honor system still largely exists.

When there is need to seek out their camels, Bedouin go out into the desert to look for them. This is not as random as might initially appear, and Bedouin can typically find their camels within a couple of days. This ability draws on two elements. First, there is the inherent, intimate Bedouin knowledge of the desert and its resources, and particularly the locations in which water and tree grazing are found in areas in the Red Sea Hills, and especially in Haimur, Umm Qareiyat, and Ungat. Second, there exists a well-informed and relatively sophisticated information network (*khabrat* in Arabic), in which travelers relate sightings of camels with particular brands, as well as other information on the distribution of ephemeral grazing, the state of trees and so on. This information is passed around Bedouin communities and is being constantly updated.

Charcoal Production

Charcoal production by Wadi Allaqi Bedouin is closely related to winter sheep grazing in the Red Sea Hills. Consequently, those households that commit labor and other resources to taking sheep to these areas are also those that engage in charcoal production. As has already been seen, it is access to camel transport that is a key determinant of whether or not this activity takes place. The ability to produce charcoal therefore further reinforces patterns of household differentiation.

The labor demands associated with sheep and charcoal production are quite complementary. While away with the sheep, Bedouin allocate time to producing charcoal at the same time as watching over the grazing. Typically, teams of three or four Bedouin will accompany a flock of 200 to 300 sheep for periods of up to two months. This provides sufficient security, company, and labor to meet both the sheep and charcoal requirements. It also permits one or two of the team to return to Wadi Allaqi every twenty days or so to collect supplies such as flour, oil, *mulukhiya* (mallow), and coffee, and to take any charcoal that has been produced back to Allaqi.

The preferred wood for charcoal is acacia, and especially *Acacia tortilis*. The harvesting of the wood is managed very carefully, as Bedouin are acutely aware of the need for good tree management for the protection and sustainable use of this resource. Only dead or dry wood is collected and used in the production of charcoal, and green and growing wood is left undisturbed. Typical production levels can work out at about five sacks per person per month, although they will be commensurately lower than this when wood has to be collected over an increasingly wide area. Overheads are negligible so charcoal realizes a reasonably good profit, especially as its production takes place at the same time as the livestock are been watched; hence the complementary labor is crucial.

In recent years, however, the relative contribution of charcoal to household income in Wadi Allaqi has been in steep decline. In the past, the strength of charcoal production has rested with its being complementary to sheep production. But this is also its weakness. With the decline, and in some cases disappearance, of transhumance associated with sheep being taken for winter grazing in the Red Sea Hills, the opportunities to produce charcoal have been much reduced or eliminated. In the same way that there are concerns about whether the winter transhumance of sheep will ever be restored when the current drought ceases, there are equal and associated concerns about the future of charcoal production as a key economic activity for Wadi Allaqi households.

Cultivation

Over the last fifteen years many Bedouin have taken cultivation increasingly seriously as part of the household economy, not as a replacement for livestock herding but as a complementary activity, and one that spreads risk somewhat more widely. Every household in Wadi Allaqi is now currently involved in cultivating small farms. These vary in size from a few square meters to a couple of hectares in some cases; the same farm can also vary in size from one year to the next. Two key factors underlie these variations, the first being the availability of labor, particular at the land preparation stage, and the second the availability of water, either soil water that might be left behind for a short period in the surface layers after inundation or by extracting water from specially dug wells. The average number of crops per farm is between five and six, but some farms grow as many as ten different crops, not only to meet dietary demands, both of humans and livestock, but also to spread the risk resulting from the possible failure of one or more crops in any season. The most commonly grown crops are watermelon, maize and wheat, followed by sorghum and tomatoes. Some households took the opportunity during the early 1990s to plant permanent tree crops, including lemon, acacia, *Balanites aegyptiaca*, and *Sesbania sesban*. However, the lake inundation to 182 meters above sea level at the end of the 1990s saw the end of these trees.

There is some disagreement among Bedouin as to the best locations for cultivation. Opinions on the best areas are split between three locations: the runnel formed by flowing water in the center of the wadi; sites at the edge of the wadi; or sites near the lakeshore. To some extent the choice of preferred site depends on the ethnic group of the respondent. In one of our surveys, for example, four out of the six Bishari respondents preferred the runnel. However, the situation was far less clear with regard to Ababda respondents; of the twenty-two, nine preferred the edge of the wadi, seven the runnel and six the lakeshore. It may be significant that the Bishari, as relatively recent arrivals in Wadi Allaqi, appear to have a rather greater consensus on the best locations for cultivation. This may be due to the fact that Bishari experience of cultivation in the desert has been rather limited, a reflection of opportunity, and where it did take place it was typically confined to runnels within wadis. The Bishari have, therefore, simply transferred previous experiences across differing environments and come to the conclusion, based on limited Allaqi experience, that the runnel still affords the best potential. On the other hand, the Ababda, with a longer history of settlement in Wadi Allaqi,

Figure 4.2: A member of the Ababda tribe watering his small cultivated plot from a shallow well. Note that the plot is protected against predators by a fishing net. Photograph by Ian Pulford.

have had the time to develop experiences based on all three alternatives, and have developed strategies to suit particular needs under differing household circumstances.

Very clear advantages are perceived of a site in the runnel located on the floor of Wadi Allaqi. Many talked about the soil being finer and hence more fertile, although some were well aware that the longer the time periods between flow events, the drier the soil becomes and the more susceptible it is to wind erosion and removal. In this context, several respondents suggested that the runnels in tributary wadis to Wadi Allaqi, where flow events may be more infrequent and more spatially concentrated on the wadi floor, are preferable to those found in Wadi Allaqi itself. However, there was a clear division of opinion about the reliability of the runnels for cultivation in the sense that water flow, which provides the rationale for cultivation, is also seen to be a major disincentive in that it can lead to the loss and destruction of crops from flooding. Some specifically mentioned the impact of torrents and suggested that this was the key reason why they chose sites away from the runnel, typically located at the wadi edge. There was the further view that as the runnels support some vegetation, however scarce, *dabuka* (camel trains) tend to follow these paths in Wadi Allaqi, and hence any cultivation is vulnerable to opportunistic camel grazing.

Of the three areas identified, the runnel is physically the most distinct, with the surface cracking of the sediment and the trapping of sediment by vegetation. Comparing the soil texture in the runnel with that in the rest of the wadi floor, the silt content rose from a value of 1 to 5 percent elsewhere on the wadi floor to between 25 and 35 percent in the runnel, and clay content from similar low values to between 12 and 15 percent; the coarse sand content fell from 50 to 80 percent to between 15 and 30 percent in the runnel. The finer texture of these soils was identified by the Bedouin as being a positive feature for cultivation, an observation that may have been supported by the greater density of vegetation in the runnel, but emphasis was placed on physical factors, such as water retention, rather than on nutrient supply, although this was recognized to some extent.

There is less clear-cut evidence for the value of the soils at the edge of the wadi. Deposits of wind-blown soil have collected at the wadi edge, but, even though these soils have a somewhat elevated content of fine sand, silt, and clay size particles, they tend to be less stable and more prone to dispersal by wind. There is also the disadvantage that rocks and stones may fall down from the hills at the side of the wadi into these areas, a feature clearly identified as being a disincentive to cultivation. Such areas are not, however, in danger of inundation by lake water or subject to the effects of flowing water in torrents, unless situated at the mouth of a tributary wadi. It appears that this security is rated more highly by some Bedouin than soil quality in the choice of cultivation site.

Lakeshore sites attracted a number of Bedouin, with virtually all respondents acknowledging the high quality of lake-derived soils and recognizing that annual inundation by Lake Nasser maintains fertility. In addition, the lake provides a reliable source of water for irrigation. However, most took the view that these sites are difficult because of the annual, and unpredictable, flooding by Lake Nasser during the late summer and autumn. "For an experimental farm the best location is the lakeshore, but for a longer-term productive farm it is better to be in the runnel," one Bedouin said. His reasoning was that the experiments were best carried out under conditions of water reliability, whereas once a rather longer term perspective was being used, then investments would be less frequently threatened by torrents in the runnel than by inundation by Lake Nasser. There may be a view, held by some, that in any case access to water is a more important determinant of cultivation site than soil quality.

Inundation by lake water, in addition to depositing a layer of lacustrine sediment, has a considerable influence on soil properties, especially on the content of soluble salts. After a short time of exposure following inundation,

Figure 4.3: Cultivation plot of Bedouin settlement on the lake shore surrounded by water during regular inundation by Lake Nasser. Photograph by Irina Springuel.

evaporation from the soil results in the concentration of salts at the surface, giving the soil a white color. Without doubt, soils with a white color are perceived by Bedouin in a highly negative way, a view that can be supported by the high conductivity of the soil. Bedouin are well aware of the problems of salinity in the area, with some considering it to be their biggest challenge. Salinity is further exacerbated by the growth of tamarisk, and the link between tamarisk cover, found especially in the lakeshore area, and high levels of salinity in the soils is well understood by Bedouin; indeed, several have specifically made mention of the fact that they prefer cultivation sites in which there is no or minimal tamarisk present.

With regard to the choice of a specific site for cultivation, it appears that the color of the soil and the ease of clearing vegetation are the two key determining factors. White soils are clearly recognized as unsuitable for cultivation and the preferred color is invariably described as yellow, with perhaps a reddish-brown tinge, although some talk about darker red soils sometimes being better, especially if found relatively near the lakeshore. Interestingly, this latter comment is not supported by soil analysis, as high conductivity values have been measured on the red soils found around the high water line, suggesting higher salinity levels in these soils.

For Bedouin, the feel of the soil seems to be nearly as important as its color. The best soils are considered to be "smooth" with a "sandy/clay" combination. Indeed, the importance of a fine sand content is widely acknowledged as "it helps drainage." Interestingly, it is the drainage that is widely seen to be important in maintaining fertility, rather than the ability of the soil to supply nutrients. This is helped by the soil being "loose." Coarse sand is to be avoided if at all possible, and soils that contain significant amounts of gravel and/or stones are definitely to be avoided; Bedouin frequently mention this without even being prompted. The reasons for such avoidance include the difficulty of working such soils, as well as their limited fertility.

Although many of the soils are high quality, they are fragile and in need of careful management. Bedouin manifest this in a number of ways, but particularly with regard to fertilizer use. Although most recognized the need for less fertilizer on clay-rich soils, there is still a view that fertilizer use is necessary to a greater or lesser extent on Allaqi soils. Indeed, some Bedouin bring clay from other areas, such as the wadi edge and the lakeshore, to increase the clay content of cultivated soils and improve fertility. The most common fertilizer source is sheep or goat dung obtained by grazing animals on land that is subsequently to be cultivated. However, this is considered to be a less than satisfactory source, especially in terms of volume but also in terms of quality: the generally poor quality of grazing material results in relatively poor-quality dung in terms of fertility. Some supplement this source of *in situ* dung by transporting in extra loads from other locations. Dung is seen as being particularly important for watermelons and maize, as well as for some tree crops, particularly lemons. Burnt vegetation is valued as a realistic alternative, with the advantage that it gives greater control over volumes applied. It is considered to be a particularly effective fertilizer for tomatoes and watermelons, as well as being a preventative measure against insect damage, a particular problem for watermelons.

The use of chemical fertilizer generates considerable discussion and disagreement among Bedouin. The only points on which there is general agreement relate to the expense of buying such fertilizer and a questioning of how cost-effective it is anyway in the context of their agriculture. A strong view expressed by those Bedouin not using such a source is that the soils are sufficiently fertile in their present form and do not require extra input, at least not from such an expensive source. There is the further view that the use of chemical fertilizer adds to salinity levels, an extra threat that cannot be tolerated.

Interestingly, the decision as to whether to use chemical fertilizer is frequently determined by whether the crop is to be sold, and especially so in the case of both watermelons and tomatoes. Most recognized that the use of chemical fertilizer produces larger fruit, but that this is very much at the expense of taste. The consequence is that chemical fertilizer is used for those crops to be sold as "people in Aswan are much more interested in quantity than quality," and other fertilizer sources are used for crops to be consumed within the household to maximize "the natural taste."

There is also a recognition that in order to be effective the use of chemical fertilizer requires access to reliable and quite large amounts of water, and herein lies a further problem. Many, but by no means all, households have access to a petrol-driven pump, but only pumps with limited capacities. Consequently, this is a key constraint on the area of land cultivated. More significant, however, is water reliability. As has already been noted, the typical annual variation in the water level of Lake Nasser of six to eight meters results in a lateral variation along the floor of Wadi Allaqi of up to ten kilometers between the highest and lowest water levels in any one year. As a result, cultivated plots that start off being located within a kilometer of the lakeshore, and hence with few difficulties of water access, can find themselves three months later being six or seven kilometers away, with all the resultant problems, not only of a limited water supply, but also of a water supply that becomes increasingly brackish. For many Bedouin this, rather than soil fertility, is the key issue.

The physical management of the soil is taken very seriously. Land clearing and fencing are the two most time-consuming and onerous tasks, and, to a degree, they are related. Vegetation that is cleared is then frequently used as fencing material. This is crucial, as growing crops need protecting from marauding sheep, goats and camels, all of which are seen as particular pests by Bedouin. Bird-scaring is another important task. Wadi Allaqi is an important bird flyway between East Africa and Europe, and crop loss from bird attack is a major issue among Bedouin cultivators. Various measures are taken, including covering crops with fishing nets or attaching old clothes, cans, and even bits of dead birds to ropes that are then strung across the plots over the growing crops.

In summary, it is possible to distinguish three broad areas of Bedouin understanding of soil characteristics that they use to support their cultivation activities. There is a good understanding of the physical characteristics of soils, especially in relation to aspects such as water retention, drainage, and erosion risk. These are all factors that are readily observed visually and therefore can easily be absorbed into the local knowledges of the Bedouin. Less easily observable, but also understood by the Bedouin, is soil fertility. Clearly understood is the value of natural materials such as animal dung and plant ash, possibly as a result of observing improved vegetation growth in soil to which these materials have been added as a result of grazing or the burning of vegetation. They are equally aware of the potential use of synthetic fertilizers, but also of their cost and their effects on food quality.

The third area is concerned with those soil characteristics about which the Bedouin are apparently unaware. This is exemplified by pH, a factor argued by 'science' to be a major controlling parameter in many soils, and a soil measurement from which much other information about a soil can be inferred. Unsurprisingly, perhaps, pH as such has never been raised by Bedouin, nor, more significantly, have they mentioned acidity or alkalinity issues. This may be because the soils of the area, although alkaline, are not overly so. It may be that what science understands as pH is simply not important for Bedouin in this setting. Alternatively, pH is not a factor that is immediately apparent by visual observation, except in cases of pH extremes where vegetation is killed off or will not grow. It may well be that Bedouin recognize that under such circumstances there is a soil problem, but it is attributed rather to high salt content rather than a high alkaline pH.

It has to be recognized that most Bedouin are using and managing soils for cultivation only because grazing and charcoal resources alone do not provide sufficient levels of income to support household reproduction. There is, indeed, an ambivalent attitude toward cultivation, with many taking the view that cultivation does not represent an economically realistic alternative to livestock herding. A common view is that it is not worth committing scarce labor resources to cultivation, when greater levels of expertise are available for sheep herding and production. However, it would be misleading to suggest that the view that cultivation is irrelevant is universally held among Bedouin; indeed, many see crops and livestock as very much complementary activities, but recognize that the soils need a significant amount of care if they are to satisfy needs and remain productive. There is also a minority view in which cultivation is actually preferable to livestock herding, especially if water availability is not an issue, in that crop production is seen as being easier to manage than that livestock. Moreover, there is the reward of "soils responding to effort," as one Bedouin put it.

Medicinal Plant Collection

Medicinal plant collection is a minor economic activity, not only in terms of the labor allocated to it but also in terms of its contribution to household income. Only a few households are involved, and then rarely on a systematic basis. Medicinal plants are usually collected on a demand basis, mainly to treat an illness or ailment within the household itself rather than for sale in the markets of Aswan or elsewhere.

It tends to be women who have a better understanding of the use of medicinal plants and other remedies, but, as with other aspects of Bedouin life, their knowledges—and how these are used—have been changing. Although the preference for modern medicine has been supplanting Bedouin reliance on traditional remedies, the medical skills held by many of the Bedouin women, especially older women, are generally valued and represent an important knowledge in the community. Indeed, some women are well known for their medical skills. For example, a medicinal plant called *kharwa‘* (*Ricinus communis*) is occasionally cultivated by some as a treatment for digestion systems and for controlling fevers. People are well-informed about plants like *Lawsonia inermis* (henna) as a treatment for burns and headaches, and lemon is used directly on scorpion stings. Others put acacia seed mixed with henna on wounds, and *harjel (Solenostemma arghel)* is used for a range of illnesses. However, care must be take with *harjel* as it is reputed not to retain its medicinal properties when grown on farms with irrigation, so many Bedouin take the view that it must always be collected from the wild.

Other Economic Activities

Over the last fifteen to twenty years, wage-labor has developed in the Wadi Allaqi area and a number of Bedouin men have found some limited local employment. In the mid-1970s deposits of economically exploitable reserves of marble and granite were discovered, mainly in the Haimur area in the hills to the east of Wadi Allaqi. The Aswan governorate established a company, Marnite, to exploit these resources in a number of quarries to the east of Wadi Allaqi. In the region about one hundred jobs have been created locally, and some Bedouin are employed as laborers, desert guides, watchmen and as other casual labor. A handful of other jobs have been created for laborers on the three South Valley University Unit for Environmental Studies and Development experimental farms in Wadi Allaqi and the adjoining Wadi Umm Ashira, as well as jobs as guides for the Allaqi field station of the Egyptian Environmental Affairs Agency. However, the scale of this employment is really quite limited overall, and it has had little discernible impact on raising household incomes in the area.

An activity that has had an important impact on some Wadi Allaqi households has been the servicing of passing camel trains *(dabuka)*. It has been estimated that about 100,000 camels per year make the journey across the desert from Abu Hamed in Sudan northwards through Wadi Gabgaba before joining Wadi Allaqi, where the camels take on some water before continuing their journey further northwards to the major camel market in Daraw, some thirty kilometers north of Aswan in the Nile Valley. The journey from Abu Hamed across the Eastern Desert is the alternative to the more famous Darb al-Arbain route on the west side of the Nile through the Western Desert and takes a minimum of ten days, although some drovers take longer, sometimes up to fourteen days, depending very much on expected market conditions at Daraw. Most *dabuka* stay in Wadi Allaqi only a few hours at the most, as the drovers are not keen to allow the camels to drink too much water from the lake, nor to eat too much fodder, as this will only make them sluggish for the remaining journey of three to four days to Daraw. Consequently, given the number of camels that come through this route each year, there is remarkably little pressure put on the grazing resources of Wadi Allaqi itself. Each *dabuka* typically comprises 300 to 400 camels, and the peak months for passing *dabuka* are between November and March.

The *dabuka* provide an economic opportunity for some Bedouin households, largely through the transport opportunities that are available. Allaqi Bedouin sell charcoal and medicinal herbs to the drovers, who then resell these goods on arrival in Daraw. The drovers are also a good and reliable source of news and information from the desert (*khabrat*, as mentioned earlier), passing on information about the locations of particular owners' camels, the location of ephemeral grazing, the condition of trees and, it has to be said, general gossip. The *dabuka* can also be a source of new camel stock to the Bedouin. New foals and sick camels are often left with Wadi Allaqi Bedouin, sometimes as payment for food and hospitality offered.

Over the last five to ten years it would seem that the importance of the *dabuka* to the Wadi Allaqi economy has declined, both in relative and absolute terms. This has come about for several reasons. First, in the past the absence of roads meant that the camel provided the most reliable means of transport, and hence the *dabuka* could provide a much-needed service for Wadi Allaqi Bedouin. Nowadays the presence of an asphalt road to within twenty kilometers of Wadi Allaqi and the greater reliability of four-wheel drive vehicles has much reduced the necessity for camels as a means of transport. The *dabuka* drovers therefore have less control. Second, with the rise in the lake level since the late 1990s the *dabuka* have taken a more easterly route that does not pass by the Bedouin settlements. Finally, the number of camels coming across the border from Sudan by this eastern route seems to have declined in recent years. This is a result not only of reduced demand in Egypt for camels, especially as a means of transport and power, but also of a major security clampdown by the Egyptian military and other security forces, especially since 11 September 2001, which has resulted in very strict border area controls.

Bedouin Indigenous Knowledges

Underpinning much of this discussion has been the central role of the local Bedouin in the management and decision-making associated with the Wadi Allaqi Biosphere Reserve. Indeed, this is wholly consistent with the original UNESCO concept, which initially used the term 'Man and Biosphere Reserve.' Although somewhat politically incorrect these days because of its gender bias, the term does, however, capture the importance of people in this process. Much of the discussion so far has drawn on both formal scientific knowledges of Wadi Allaqi, as evidenced by the work led by the Unit of Environmental Studies and Development of South Valley University in Aswan, and the knowledges held by the Bedouin themselves about the physical environment in which they find themselves and from which they must make a living. However, making use of this indigenous, traditional, and local knowledge system can be no less problematical than making use of formal science in trying to understand environment and development issues.

There has arisen a view among some development practitioners, rather more implicit than explicit, that where indigenous environmental knowledges are seen to be present and held by members of a community, they should be captured and used in the promotion of sustainable development. The trick is how to find and use these knowledges. There has developed a tendency in such approaches to see indigenous knowledge as something that is unchanging, even static and timeless. However, the Bedouin of Wadi Allaqi demonstrate that this is not the case at all, and that their local knowledge repertoires are in reality dynamic, provisional, transitory, and highly negotiable. The Bedouin constantly rework and reevaluate their knowledges, constantly adding new ideas and experience, and are more than prepared to discard existing ideas that have been superseded or become irrelevant. Environmental knowledge is constantly acquired, tested and reworked, even if this is only to confirm what is already known. Second-hand information is not to be trusted without being first tested and experienced. First-hand experience produces knowledge that an individual never forgets, one Bedouin argued; another remarked: "Environmental knowledge is about observation, experiment, and only then knowledge." Secure knowledge about preferred grazing species is frequently developed from observation of animals' feeding behavior and subsequent growth rates. Such a reworking and reevaluation of knowledge is a slow and careful process. It has to be. For Bedouin living close to the margins, they have rather more to lose if they get things wrong.

The Bedouin also demonstrate a lively dynamism in their development of local knowledge. For example, Bedouin have observed that their small livestock, particularly sheep, prefer crop residues to other feeds. As an experiment, a Bedouin woman fed her livestock a range of cultivated feeds, the results of which showed that her livestock preferred *bersim hijazi* (lucerne) rather than another variety, *bersim baladi*, as the latter contained a higher proportion of water in relation to feed matter. Significantly, because it was a woman who had led these experiments, this resulted in a degree of women's empowerment within the community. Indeed, this empowerment was subsequently further reinforced by the use of *shilbeika*, an aquatic plant found in Lake Nasser and unknown in Allaqi before the construction of the Aswan

High Dam, for animal feed. After sheep were observed eating dried *shilbeika* washed up on the lake shoreline, some women collected it, dried it, and fed it to their sheep, an experiment that was successful and generated new ideas and new knowledge. The current use of *shilbeika* is seen to be very much the product of women's efforts and, indeed, for many men the suitability of different varieties of the plant, and how much they should be dried before being fed to sheep, is still largely unknown.

Implicit in the romanticizing of indigenous knowledges is the sense that indigenous environmental management is somehow ecologically harmonious, and that indigenous resource management will necessarily promote sustainable development. While this may often be the case, it is not always so, as groups must sometimes take action to facilitate their own survival. In Allaqi, some changes are forced, such as the restrictions on winter hill-grazing resulting from several consecutive years of drought. A consequence has been the emergence of extreme grazing pressure in parts of Wadi Allaqi, as there are no alternative sources of grazing available. There is also the view that indigenous knowledge is closely associated with identity. While this may be the case in the most general of terms here in the wadi, the Bedouin are highly pragmatic about the knowledges they hold. If any agricultural or environmental idea has been supplanted, or is of no practical use any more, it is quickly discarded. Clearly, indigenous knowledge is adapted to deal with new opportunities and new resources, and new, hybrid knowledges constantly emerge and develop. For indigenous knowledge to have value for the Bedouin of Wadi Allaqi, it must contribute to the enhancement of production, a markedly utilitarian view.

The dynamism of economic activity ensures that knowledge must adapt to changing economic circumstances and opportunities. One of the Wadi Allaqi Bedouin was well aware of the fact that his environmental knowledge repertoire, and those of his two sons, was changing and adapting to new circumstances. His sons now know virtually nothing about the natural environmental and its resources beyond Wadi Allaqi itself, because the household had chosen not to migrate to the Red Sea Hills for winter grazing for several years. Thus, the sons have not had the opportunity to observe, evaluate, and work with the different plants and grazing resources found in these areas. Even the knowledge base of the man himself was being eroded through lack of engagement. Two reasons explain this. First, the drought that has affected the Eastern Desert since the late 1990s has discouraged migration and encouraged a greater use and development of alternative shoreline grazing resources along Lake Nasser in Wadi Allaqi. The longer that these winter migrations fail to take place, the more likely that the environmental knowledges the Bedouin retain of such areas will gradually disappear. This is a particular risk for younger people for whom these areas are now becoming economically irrelevant. Already, at least one family has become quite content with the more sedentary life in Wadi Allaqi, and the advantages of easy access to lake water and the Aswan market. This family talks about building a permanent home, at which point a return to seasonal migration patterns elsewhere in the desert will be even less probable. Knowledge of these grazing resources is then likely to be lost forever.

Second, with the major economic changes and opportunities that have been caused by the creation of Lake Nasser, new local knowledges have had to be developed. Consequently, there has developed among the younger generation something of a knowledge vacuum concerning traditional grazing areas and other resources in the wider desert, such that there has developed a 'generation gap,' with older Bedouin possessing a different range of environmental knowledge than their sons. For example, one Bedouin commented that his son had not had the opportunity to learn the direct route from Sayalla to Umm Ashira (two important locations in the desert) because "the days of camel travel were over." Pick-up trucks are the new camels, and his son's knowledge reflected this; as his father dryly put it: "He can tell you the owner of any passing pick-up . . . and as long as there is a track he will not get lost." But other changes have contributed to this; for instance, as market activities have become increasingly important for Wadi Allaqi Bedouin over the last ten to fifteen years, so there has developed a marked tendency to rework and evaluate environmental knowledge within households on a more active, regular, and systematic basis to meet these changing demands.

For the Bedouin, if indigenous knowledge is to contribute to sustainable development then it has to be anchored within the economic and socio-cultural structures of the households involved. It has been made clear that there has developed quite a marked wealth differentiation between Wadi Allaqi households. This issue reemerges with regard to indigenous knowledge acquisition and development. Those households with the wealth to afford camels exhibit a wider indigenous knowledge base than those who do not. Those owning a sufficient number of camels, for example, have the resources and capacity to take livestock for grazing during the winter months in the Red Sea Hills or on the lakeshore crop residues. Such access is unavailable to poorer families, who do not have the same the economic resources in the form of camels to travel to these areas, and therefore their knowledge bases of these areas are nothing like as informed as those of the wealthier Bedouin.

However, there is more to this than just the economic. Social and cultural contexts are also highly significant. Sedentarization is creating new social roles within the family and thus also different knowledges. Knowledge acquisition is, to a large extent, controlled by experiences that accrue with age and, due to the different spatial extents of men and women, gender. The concept of gender is used here carefully, recognizing that there are different groups of men and women in the household whose knowledge and decision-making power are differentiated. Within those more sedentary Bedouin families, the extent of women's environmental knowledges seems to be changing rather more than those of the men. Women's work, and the knowledge required for it, is determined by their location. If the resources needed for household reproduction are located close to the household itself, then women will have developed a knowledge of them; if resources are located further afield, in more overtly male spaces, women's knowledge is much less well-formed, or even non-existent. This impacts not only on women's knowledges, but also on those of the entire family. Women are responsible for the early education of children up to the age of ten or twelve years, which means that women's more restricted environmental knowledges tend to reduce children's early environmental education in scope. Thus, sedentarization, and the greater involvement with market economies that has followed, has meant that some women have seen their roles as environmental knowledge experts actually reduced, notwithstanding the comments made above about some other women feeling more empowered through successful grazing experimentation.

It is clear from the evidence of the Wadi Allaqi Bedouin that there is no such thing as a pristine indigenous knowledge. Indeed, a key part of the process involved in the development and evolution of indigenous knowledge is the acquisition and evaluation of knowledges from outside Wadi Allaqi. Outside information and knowledge filters in. This may be through visitors to the area; return migrants who have spent extensive periods of time away from the area but have now returned on a permanent basis; visits made by Bedouin from the area to urban markets such as at Aswan; Bedouin individuals or groups who arrive from different areas; and so on. In other words, there exists a range of different sources of information, all of which may be evaluated in different ways, and to greater or lesser extents, by the Bedouin.

Although family and kinship contacts are crucial to environmental knowledge acquisition, external, non-family sources of information can be equally crucial. Many Bedouin talk about acquiring information through friends, neighbors, and acquaintances throughout their adult life, and about exchanging information and knowledge, sometimes on a daily basis. Wadi Allaqi Bedouin women talk about discussions with Nubian women, whom they meet when visiting Nile Valley relatives. Such meetings have resulted in information about seeds, types of soil, and planting and harvesting times being absorbed by Bedouin, something that has become increasingly important with the process of sedentarization that is currently taking place in the area. Bedouin men spend considerable periods of time talking with a range of 'outsiders,' including incomer lakeshore farmers in Wadi Allaqi itself, Nile Valley farmers during the Bedouin's increasingly frequent visits to Aswan, and men passing through Allaqi. Consequently, what development planners might see as an indigenous knowledge in fact is not. What they are confronted with is a mediated, local knowledge that comprises a hybrid of various knowledge sources that are evaluated, reworked and deployed in the interests of household reproduction. Hence, Bedouin 'indigenous' knowledge is not something simply internal to this group, but has long been influenced by contact with various external groups. This clearly poses a problem, because all too often indigenous knowledge has been character-ized as an alternative to scientific knowledge, whereas in fact local, so-called indigenous knowledges are more than happy to acquire, appropriate and use knowledges that are firmly grounded in the traditions of Western sci-ence, if those knowledges make economic, environmental, and sociocultural sense to the community. In a similar way, there is no such thing as shared community, household, or family knowledge. There is an unevenness in the knowledges held across individuals; some people's knowledge is different, limited and/or partial, compared with knowledges held by others. What has become apparent from living with the Bedouin in Wadi Allaqi is that there are multiple environmental knowledges. This does not mean that people necessarily hold radically different, conflicting, and opposing knowledges about the environment, but that they typically retain different emphases with regard to that knowledge and how it is subsequently deployed. The content and depth of environmental knowledge is uneven across both com-munities and households, suggesting that the concept of a (singular) com-munity knowledge is unhelpful.

These are all very difficult and thorny issues, but ones that cannot be avoided. There is no doubt that the Biosphere Reserve concept offers real hope and opportunity by involving local people in its development and management, but there has to be a recognition of the ways in which the knowledge systems of the Bedouin in Wadi Allaqi can genuinely contribute. This is not to suggest that indigenous Bedouin knowledges are in some way superior to knowledges produced by Western, 'rational' science, rather that they offer another way of thinking about and conceptualizing the natural environment in the interests of sustainable development.

Summary

There is little doubt that the Bedouin of Wadi Allaqi have taken the opportunities offered by the new natural resource opportunities, although it is apparent that some households have been able to access some of these resources more successfully than others. Indeed, those households that have been able to do so have become wealthier, and economic differentiation between households in Wadi Allaqi appears to be widening. The economic activities of the area are, nonetheless, still dominated by small livestock production, particularly of sheep, although for a few households, and again it tends to be the already wealthier ones, camel herding is still important. Increasingly, however, cultivation continues to gain in importance, and especially as the communities become more and more sedentarized. Interestingly, it appears to be the women of the community who are embracing the resource opportunities of Wadi Allaqi rather more than many of the men. Whereas many men still cling to the idea of being livestock producers and herders, even to the extent of still retaining geographical ties to the desert, although this has been interrupted in recent years by the drought, many Bedouin women are more interested in developing the resources of Wadi Allaqi, including cultivation opportunities. This next chapter, therefore, goes on to explore the changing role of women in Wadi Allaqi and gender relations more generally in more detail.

5
Bedouin Culture: Stability and Change[1]

There have been significant changes to men's and women's household roles as a result of increasing Bedouin sedentarization with the formation of Lake Nasser. While men's lives have continued to be focused around the grazing and marketing of sheep, and hence follow an extensive spatial pattern, women no longer pursue this nomadic lifestyle and instead remain around the shores of Lake Nasser throughout the year, moving short distances only to follow the seasonal movement of the lakeshore.

One of the most noticeable outcomes of this sedentarization has been the introduction of agriculture, initially on a small scale and run by women, but on an increasingly large scale and now involving the entire household. This, plus the increasing amount of contact between Bedouin and outside communities who have come to Wadi Allaqi along the new asphalt road to fish, mine, and farm, has meant the appearance of different opportunities for men and women and thus has changed, sometimes quite subtly, the nature of gender relations.

Bedouin Households in Wadi Allaqi

On the whole the division of gender roles in Wadi Allaqi is based on scale and formality. Men's roles involve large-scale decision-making about the use of resources and household activities, while women are concerned with more 'mundane' decisions on a day-to-day basis. Similarly, men dominate decisions about resources that are exchanged for commercial value, and other household decisions that are 'political' in nature, while women's decisions are generally confined to subsistence issues. However, despite the insistence of a number of Bedouin men that they, or the head of household, alone made decisions, there was evidence that, in practice, decisions were taken in negotiation or discussion with women and the remainder of the household. The gender division of decision-making is thus not the simple binary that it might first appear to outsiders, but is arranged around patriarchal authority. Not all men have freedom in their decision-making and may have to defer to the authority of the head of household. In practice, some men may not have a great deal more autonomy in decision-making than women. Thus, there is a quite complex system of decision-making responsibilities in the Bedouin community in the wadi. However, it would seem that women's roles in the management of particular resources, and thus their knowledge of them, depends on their proximity to these resources and thus the women's ability to be involved with their use.

Women take the lead in important everyday areas of daily decision-making, even when the men are around. For instance, when households are established at the lakeside, women have a decision-making role in the management of water resources for household supply. They have to manage the labor (of women and children) to bring sufficient supplies to the household each day. In addition, women and older children have day-to-day autonomy over managing the small flocks of sheep, usually when they are sick or pregnant. It is still men who manage flocks of more than forty animals, and so it is the men who make decisions about the overall management of the household's livestock. In addition, when men are absent from the household during the winter, when they take animals to find distant grazing, or are conducting business in Aswan, women have to take household decisions. In this case, women do have more responsibility (although, in some cases, young sons take over their father's responsibilities in his absence). Owing to this absence of males, women have a great deal of knowledge about managing their household resources, including livestock, especially at the local scale.

As a result of the traditional association of Bedouin men with animal husbandry in Wadi Allaqi, it is women who tend to make decisions about agricultural production—if this activity is engaged in at all. In part, this is due to the fact that men spend long periods of time away from the household, and also that it is not always seen as an activity appropriate for Bedouin men, whose sense of identity is very much tied up with the experience of long desert expeditions. Some women also feel that this is a skill that is particularly female. Many Bedouin women think that men are not particularly good at managing the farms because they do not share the same knowledge of soil, irrigation, and plant types that the women do. Much of the women's knowledge had come in the past from talking with Nubian neighbors, or from discussion with UESD researchers visiting Wadi Allaqi. However, as agricultural production has become more significant within the household economies of Wadi Allaqi Bedouin, men's influence in agricultural decision-making has become more established (as will be discussed below).

There are many different and conflicting demands on women's time. What is noticeable about women's work is the extent to which they carry out a number of tasks simultaneously, such as carrying babies, cooking, and supervising those children looking after animals. Especially before the adoption of agriculture, the greatest demand on women's time was the collection of plants for feeding sheep and goats. A range of different plants was collected, some of which needed to come from the lake, such as *shilbeika* (aquatic plants), and so required the additional effort of wading through the water. Aquatic plants need to be dried before they are given to sheep and goats, and at times of high plant availability women dried and stored fodder to give to the animals in times of scarcity.

Cooking is also a high priority and can involve one or two sessions per day, in addition to the daily task of bread-making. Hot stews, including *ful* (beans), *'ads* (lentils) and *fatta* (bread, meat broth, and meat), are common in winter when the weather is cool, while *'asida* (milk, fat, flour, and water mixed into a paste) is eaten all year round. There is a particular demand for vegetables during Ramadan, when there is a desire to prepare something special for the evening meal *(iftar)* at the break of each day's fast at sunset. Women's cooking work is rather less during Ramadan, as they only need to prepare a simple, but filling, early morning meal *(suhur)* before dawn (often lentils), and then prepare *iftar* for sunset.

Women also collect wood for cooking. Most are of the view that tamarisk gives a good fire (although acacia wood is seen to give the best fire), but when this is in short supply (for example, when tamarisk is too wet owing to inundation), women use dried animal dung, old cloth and other scraps, and waste material to supplement the wood. Kerosene stoves *(babur)* are not commonly used by Bedouin because of the prices of kerosene. When such stoves are available, they are thus used only when absolutely necessary.

Water has to be collected by women for cooking, drinking, and the irrigation of farms. Owing to fluctuations in the vertical levels of the lake over the year of up to six meters, this can be an extremely onerous task at certain times, particularly when the water is retreating in spring and summer. Children often assist in this task. This water is used for general household use. The introduction of agriculture intensified this task until the use of mechanically driven pumps became widespread. Farms require large amounts of water and, as a result of lake level fluctuations, water must be carried over large distances five or six times a day. This effort initially, and not unreasonably, has dissuaded many women from taking up cultivation, even on a very limited scale, and this is clearly a key problem in promoting the expansion of fodder cultivated for livestock.

Women also have a crucial pastoral role in looking after both children and goats and sheep. In terms of childcare, location is important. Hence, many young women may establish their household near to that of their mother after marriage. This has several advantages. Mothers can help specifically with childcare, as well as contributing more generally to the extended family. This proximity provides security, especially while husbands, fathers and brothers are away from Wadi Allaqi. It also means that there are likely to be some men or older boys around who can undertake tasks that women are prohibited from doing for cultural reasons.

Women generally have the responsibility for looking after young and sick sheep and goats, and a small number of mature animals in the local environment. When the majority of the flock is away, some animals are left behind with the women to supply milk for those household members remaining in the wadi. On the whole, women supervise children, especially girls, who look after the animals, but there are times when women become more directly involved in watching the livestock.

The final significant task that women undertake is moving homes from the threat of inundation, or toward the lakeshore when water levels retreat significantly. Although the decision to move is primarily a male one, men may be absent grazing sheep, doing business in Aswan or making charcoal. Hence, women may have to make the decision when to move and then have to do it themselves. The time of moving is dictated fundamentally by the lake-level, and so if men are not around it has to be the women who take charge of the move. This involves a huge amount of work to pack and then move tent materials, cooking equipment, possessions, children, chickens and so on.

Women's spaces are located in the immediate area of the family household. Some fifteen or twenty years ago, when visitors arrived at Bedouin encampments men were met at the visitor's area located at some distance, typically up to 100 meters, from the remainder of the household. As a result of increased familiarity with visitors as more people come to Wadi Allaqi (particularly after the construction of the asphalt road), the distance between the guest area and the main encampment of a family has been steadily reduced over recent years, and now may be at a distance of only twenty or thirty meters. Nevertheless, while female visitors may now be taken to the domestic part of the household in the main encampment, male guests are still met and entertained in the specified visitor's area. This clear geographical distinction unequivocally marks out women's and men's spaces, with women's spaces very much being those of reproduction, based around cooking and childcare responsibilities.

Women appear to be happier than men with an increased sedentary life. Permanent residence sites have the advantages of being nearer reliable supplies of water, of providing space where children and small animals can be looked after in a relatively safe environment, and where women themselves can benefit from the support of other families. The last point is important because such support ensures that in times when most or all of the family's men are away with their sheep, there will be some other men and older boys around to help with some of the heavier work and those tasks that women are culturally forbidden from undertaking, such as milking animals. Furthermore, sedentarization allows for cultivation to take place, even on a relatively modest scale, to supplement both animal feed and household diet.

Whereas women generally have less enthusiasm for a more mobile nomadic lifestyle, and are quite happy to benefit from the advantages of a sedentary lakeshore location, many men indicated in discussions that mobility is an important aspect of their life and, indeed, self-image. It is important to male Bedouin self-identity that sheep and camels are grazed in the desert and that Bedouin men are part of this process. Being at one with the desert is central to being a Bedouin. It is something that Bedouin women reflect on with affection: "Let the men go to the desert and play at being boys again—we are happy to stay here."

Particularly during periods when their menfolk are absent, women's spaces are not just of the household, but of the local environment. They are the managers of local resources when men are away seeking better grazing with the majority of the livestock, when they are making charcoal, or are doing business in Aswan. Women have to supervise animal grazing and collect water and materials for fodder, as men can be absent for many weeks at a time. It would seem that their workload increases significantly when men are absent. However, there is a belief among women that men's presence or absence does not make much difference because women do most of the work anyway.

These distinctive gender roles are both reflected in the arrangement of space and movement in Bedouin settlements and are themselves reinforced by the spatial form of the settlements. Most directly put, there are different spheres of work. On the whole the men have the greater access to the spatial reaches of the desert, other groups of Bedouin, and market places. Women tend to be constrained to the domestic—the space of the household and its immediate environment—although this is a variable concept both in spatial extent and in that it changes at different times. While men take animals to distant grazing resources, the small numbers of animals that are left with the women are confined to grazing around the household.

New people in Wadi Allaqi provide opportunities for Bedouin. Most significant has been the establishment of commercial farms around the lakeshore, run by farmers from the Nile Valley, as discussed earlier. Farmers allow Bedouin to graze their animals on the residue after harvest, a mutually advantageous arrangement as the Bedouins' sheep and goats fertilize the fields. Despite this change, there is still a sense of a domestic space, generally around the area of the encampment, where women can tend their animals and are unlikely to bump into unfamiliar men. Bedouin women in Wadi Allaqi perceive that they have to remain very close to the household and that married women must have a male escort if it they want to take animals to graze on agricultural residue. Because of this, in practice women simply do not go to agricultural areas: the fields are considered a distinct social space forbidden to women. There are a number of reasons for women's exclusion from the agricultural areas. Most importantly, it is *'eib*, or shameful, for Bedouin women to encounter farm laborers. Furthermore, it is the men who must negotiate grazing access with the farmers, and it is thought that women might not be able to prevent animals from exceeding agreed grazing or from damaging nearby crops. As a result of this gendering of space, Bedouin women do not value agricultural residues as an important resource to the same extent as the men. The women of Umm Ashira, for example, have no knowledge of agricultural-residue grazing, as this is a male activity undertaken far from where the women live. Before the widespread adoption of agriculture by Bedouin, they perceived lakeshore grazing to be the best grazing for sheep and goats, although they also recognized the seasonality that affects the availability of plants in different areas. In the summer, for instance, when the lake level is at its lowest, dense vegetation grows on the lakeshore, making this the favored grazing location. Sadinab women, whose families have no means of travel to exploit crop-residue grazing, also do not know anything about which types of residues might be best for animals. This does not mean they have no understanding of these new opportunities or the threats they might also pose. Many women are worried about their animals being poisoned by the chemical fertilizers and pesticides used by the farmers. Women's knowledge of the environment then is linked to the spatial extent of their movement within it.

This gendered geography is made more complex through the age differentiations. Young girls have freedom of movement circumscribed only by fears of their straying from the household and becoming lost. It seems that until about the age of ten or twelve, girls and boys share the same space. While the tasks they are taught may be different in some respects, there are common tasks, most important of which being the herding of small numbers of animals around the households. It is only after the age of twelve or so that the movement of girls is restricted, and that it is *'eib* for them to take animals to graze on farm residues. At this point gendered roles become total, with boys accompanying their male relatives herding animals far from the household, and girls taking on the responsibility for domestic tasks around the household.

Older women and widows have different gender roles than their reproductive sisters, and so their movements are differently circumscribed. Older unmarried or widowed women can take animals to graze on farm residues, for instance, while married women cannot, or must go with a man from her own family. It would also seem that widowed women can travel more widely and adopt a greater range of activities than married women. However, this greater freedom in terms of gender role is not something that elicits envy from other Bedouin women, rather a sense of pity that the widow is forced to take such actions in the absence of any male relative.

Women and the Household Economy

As we have seen, the Bedouin household economy in Wadi Allaqi is dominated by sheep and goat production. On the surface this looks like a predominantly male activity, as it is the men who take the animals to find the best grazing in the winter months, and who claim to know the best ways of curing sick animals. Although women may own livestock (from weddings and gifts), they generally do not make decisions about its management. A frequent comment by Bedouin women is that they are too busy with other tasks to look after sheep. In addition, there are cultural limits to women's involvement with livestock. For example, in contrast to most nomadic pastoral societies, in Wadi Allaqi women cannot milk animals, nor go to the market to sell them directly. However, it is clear that women value sheep very highly. They need animals around them to supply milk for daily use, and they look after young animals (particularly kid-goats, which tend not to follow their mothers as closely as lambs) around the household.

Perhaps more importantly, there are clear cultural reasons for having animals, particularly sheep, as part of the household. The presence of sheep in a household means that it is blessed *(baraka)*. One Bedouin woman from Wadi Umm Ashira said, "If a woman has nothing but one sheep, she is still a rich woman." Quite literally, sheep means wealth: the Arabic word for rich *(ghani)* may be related to the word meaning sheep *(ghanam)*. Sheep also bring religious blessing. This is best illustrated in a story told by Bedouin women of Muhammad, the prophet, who was being chased by people who wanted to kill him. When he tried to hide behind a herd of goats and make his escape, the goats spread out and left him exposed. However, when he hid behind a flock of sheep, they stood tightly around him hiding him from his enemies.

Women themselves own sheep and goats that are kept with the household flock, but which will be divided equally with the husband in the case of divorce (or split three ways if there are children), unlike a man's animals, which will remain solely his property. There are three sources of sheep and goats for women. First, they can inherit animals from their parents, although under Islamic law sons inherit twice as many as daughters. Second, sometimes gifts of a small sheep or goat can be received from a father or brother. Finally, women can receive animals on marriage. Hamidab women in Umm Ashira, for example, traditionally on marriage receive two camels, six sheep and six goats from their future husband's family. Poorer Sadinab families may only be able to offer one or two camels and a few sheep, while for Fashikab marriages the number of animals involved depends on the perceived wealth of the husband-to-be.

As a result of being limited spatially to the local environment before the introduction of agriculture in Wadi Allaqi, the quality of women's animals (and those of poorer families) is limited by the immediate environment and its resources. Animals are often of poor quality as a result of overgrazing in the environment immediately around the household. In such cases there is little or no time for plants to recover because animals are there for long periods, rather than being moved on when resources became stretched. Many fodder plants are dangerous to small livestock if eaten wet, or in large quantities. Although women know this, there are times of the year when there is no option but to feed sheep and goats with plants that have not been thoroughly dried, or to feed them only one type of plant, which has implications for the quality of their diet. With limited feed available, some women choose a cyclical use of such resources. In other words, they feed their animals only *shilbeika (Najas* sp.) or *toroba (Glinus)*, for example, until such a time that the quality of the animals clearly declines. At this point the sheep are given corn, bread, or *dura* (sorghum), until their condition recovers. At this point the animals are put back on *shilbeika* or *toroba*. This cycle is inevitably not good for animal quality, leading to poor quality milk for family consumption and less meat bulk, resulting in a lower price at the market should they wish to sell the animals.

It is clear that women have developed a deep environmental knowledge of the different types of grazing available in their immediate surroundings. Their knowledge of the plants available is as follows:

1. *Grass (hashish)* is used as a general term referring to the most favored grazing, and includes not only grass, but also other small herbs for which Bedouin women have no particular name.

2. *Glinus* (*toroba* or *hashish al-agrab* [scorpion grass]) can be grazed *in situ* on the immediate lakeshore immediately after the lake has retreated, but is also collected and dried as it can be somewhat bitter when wet. It is always new *Glinus* that is grazed or collected, because when it matures it becomes increasingly bitter and poisonous for sheep because of the higher content of toxic components in the plant's mature stage. Even fresh *toroba* can be problematic when wet, as it can cause diarrhea. When *toroba* is cut or dies, and is exposed to the sun, the toxic components decompose. Sheep and goats will eat mature *toroba* if they are hungry, as a last resort, but this can cause excessive thirst. The result can be that animals will drink too much and this in turn can cause respiratory problems. One Fashikab family estimated that they lost about 100 sheep at one time after they ate mature *toroba*. It should be noted that the availability of *toroba* is not consistent throughout Wadi Allaqi. Although it grows profusely in Wadi Umm Ashira, and also in the Sadinab area, it is less common in Wadi Quleib, although in recent years this has become less the case.

3. *Najas* spp. (*shilbeika*). This aquatic vegetation is harvested out of the lake. After collection it is dried, because if it is not it can cause diarrhea among sheep. Bedouin women are well aware of this problem. However, they sometimes feed sheep with small amounts of *shilbeika* in a wet condition if they have no dry material available as an alternative. This is an especially important issue for the Sadinab, who are particularly dependent upon *shilbeika* as a grazing resource. *Shilbeika* is collected all

year round, even when plenty of alternative grazing is available, and it is dried before being stored on the tops of tents for use in the drought season. It is important that the *shilbeika* is grazed or cut, because when the lake retreats the exposed plant that is not used rots and smells. This can prevent new grasses from coming through, thus restricting new grazing areas from developing when the water retreats. On the other hand there is evidence to suggest that *shilbeika* can help to retain soil moisture for short periods before it rots down so, ironically, helping to promote the early growth of grass.

4. Tamarisk (*abl* or *tarfa*). Only the young shoots of this plant are collected, as the older, more mature material is too salty. For this reason, wherever possible fodder is collected from inundated trees in the lake and fed to sheep, although some believe that this worsens the taste of the milk and only feed tamarisk to camels. An additional problem of tamarisk growth is that it creates shade and so reduces the amount and quality of undergrowth that could be used for grazing. Even more importantly, decayed leaf-drop from tamarisk increases the salt content of the area immediately surrounding the tree or bush.

5. Acacia leaves (generic name: *sant*; the preferred types in the downstream part of Wadi Allaqi are *Acacia ehrenbergiana* [*salam*] and *Acacia tortilis* subsp. *raddiana* [*sayyal*]). As with tamarisk, it is the fresh shoots that are preferred as grazing material. This can supply plenty of green material with a limited number of thorns.

6. *Nagila*. This is the name given by Bedouin to four grasses and small herbs growing by the lakeshore: *Cyperus pygmaeus* (small herb); *Crypsis schoenoides* (short grasses); *Fimbristylis bisumbellata* (small herb); and *Eragrostis aegyptiaca* (short grasses). The Bedouin do not distinguish between these individual plants. The grazing of *nagila* improves sheep quality appreciably, and inevitably a common complaint is that there is never enough of it available.

7. *Handal (Citrullus colocynthis)*. These are groundcover plants with melon-like fruit that are full of seeds that can be dried and eaten by livestock. Some Bedouin also cultivate *handal*; if so, it is then called *gawirma*. It does, however, have a relatively limited use as fodder because it can be quite toxic.

Bedouin girls and boys supervise sheep and goats grazing in the local environment around the household. Among the Sadinab, older boys and men also take (and carry one at a time, if necessary) sheep and some goats across to small islands in Lake Nasser. The water is relatively shallow near their settlement, and this allows the Sadinab to use the better quality grazing on the islands. The islands offer better grazing because they are untouched for long periods owing to isolation. Use of the islands also permits easier control over sheep. Pregnant ewes are not taken (even though the better grazing might be beneficial for them) because of a combination of their being too heavy and the fact that such movements are not good for pregnancy. All the animals are brought back to the household each evening.

According to the men, lakeshore vegetation when first exposed is left untouched for about two weeks to ensure that the grasses have time to thicken (but before the sun dries off the soil and burns the plants). It is then thoroughly grazed before animals are allowed to move on to the next emergent grazing following from lake retreat. It would seem, however, that this may be what happens in an ideal situation and is not always adhered to. Women explain that children do not wait for any length of time before letting sheep and goats graze on any exposed vegetation, and hence the idea of some form of controlled grazing system may not hold. Conversely, among other groups there is evidence that there is indeed a differential management of exposed vegetation. Some Sadinab, for example, suggest that their sheep are left to graze *nagila* by the lakeshore immediately following lake retreat, but the exposed *shilbeika* is actively collected, dried and then stored for winter use when the lakeshore grazing is all covered by the annual lake flood.

If grazing and/or feed are in short supply, sheep will get priority over other livestock. This may be due to cultural reasons related to the *baraka* that sheep bring to a household, but the Bedouin women also have a number of practical reasons for this. First, sheep are perceived to give a better meat return than goats on the same volume of feed. Second, goats are more prepared to roam around in search of food, and are also more tolerant of a wider range of foods. Finally, if Bedouin are intent on selling any animals, sheep generally command higher prices than goats.

In general, a prevalent view among most Bedouin women is that sheep and goats were healthier in the past, and as a consequence they produced both more and better quality milk. A number of reasons are given for this. Sadinab women believe that in the past there was a wider range of plants providing a variety in diet. Nowadays, by the lakeshore in downstream Wadi Allaqi livestock have become too dependent on tamarisk leaves. Women remember the availability of good grazing in upstream Wadi Allaqi in the past (particularly around Eigat), from where sheep returned in a generally healthy condition. It is generally believed that the grazing was so much better then because it rained more. This also demonstrates clearly the changes in environmental knowledges held by women as a result of sedentarization. As the young women no longer go to these upstream areas, they now lack the first-hand knowledge of the environmental resources of these areas. For them, there is no experience of an environment or a set of resources other than those located in the immediate vicinity of the household.

Hamidab and Fashikab women also talked about the pre-dam days, when the Bedouin had reciprocal relationships with Nubian farmers. Nubians cultivated areas in the Nile Valley on which the Bedouin would sometimes share farm work and, in return, after harvesting, the Nubians would allow them to graze their sheep and goats on the crop residues. After the dam was built and Lake Nasser was formed, the Nubians were resettled elsewhere (mainly Kom Ombo and Khashm al-Girba), and so this important source of livestock grazing was lost. It seems that some Bedouin could also have been resettled with their Nubian neighbors and friends in these two areas. However, most decided against this because the government ruled that sheep and goats belonging to Nubians and Bedouin were diseased, and so could not be taken to the new locations. A number of Bedouin, therefore, quickly took their animals into the hills and disappeared. Sadinab people told us that the Hamidab were more affected by the dam than they were, because the Sadinab had always lived in the desert and had had fewer economic and social relationships with Nubians. Hamidab, on the other hand, lived relatively near to the river, among Nubians, with the result that the dam introduced significant changes to their livelihood patterns.

A third important constraint concerns the workloads experienced by women in the area. Looking after sheep and goats constitutes only one of many activities. In addition, there are responsibilities for caring for children and ill and very elderly relatives, cooking, cleaning, water and firewood collection, and even the relocation of tents in response to lakeshore movements. The relative importance of all these activities varies between households, and, indeed, between different time periods in the same household. Sheep and goat tending is rarely consistently given top priority. Importantly, girls help out with these various household activities when they are old enough; indeed, evidence from this study suggests that one of the main areas in which both girls and boys help out with household reproduction is in looking after sheep and goats on a daily basis. For boys, this is seen as part of the growing-up process in terms of learning how to care for small animals in anticipation of their joining adult men in taking sheep to the hills for winter grazing, once they themselves reach the age of fourteen or fifteen years or so. However, children are not always as careful or circumspect as they might be in their care of sheep. In particular, little management of grazing appears to be demonstrated, and some Bedouin suggest that the grazing pressures around the lakeshore may be exacerbated by the relative lack of adult super-vision. An inevitable consequence is that, under these circumstances, sheep and goats are likely to be less productive than those grazed elsewhere away from downstream Wadi Allaqi.

A key issue that clearly constrains women's ability to improve sheep qual-ity is that of obtaining reliable and good-quality livestock feed. Naturally available grazing is neither profuse nor always present in downstream Wadi Allaqi. Equally, many women recognize the benefits to be gained from diver-sifying feed types for livestock. Because women are increasingly geographi-cally constrained to Wadi Allaqi, their environmental knowledges of this area have become deeper. Many have a firmer grasp of the area's opportunities than do the men, and particularly of the constraints for management and production. Once again, it would appear that marital status is important; although women with husbands alive rarely go to the desert, they, nonethe-less, have access to the productive capacities of the desert through their hus-bands. Widows do not have this option, and therefore have to make the best of the natural resource base available to them in the downstream part of the wadi. Widows are in an ambiguous position in terms of their gender roles in that they are forced from necessity to adopt tasks normally only available to men. Widows in Wadi Allaqi have been more interested and engaged in cultivation for much longer than other women, perhaps owing to the neces-sity of supplementing their household incomes, in addition to the greater freedom they face in undertaking productive activities.

For these reasons, perhaps, it would seem that generally women have been more interested than men in cultivation, and more so widows and poor women than married or relatively wealthy women. Women are per-haps developing a different construction of knowledge in their immediate environment, a knowledge that sees the cultivation of small farms as an additional way of exploiting local resources and, indeed, of spreading risk. It is more than this, however. Because women are now virtually permanent residents of the downstream area, they are on hand to see through the cultivation cycle. In addition, some men perceive cultivation as not being "work for a Bedouin [man]." All this would seem to point to women, and especially widows, being a group more open to cultivating feed crops in support of livestock.

One of the greatest changes on the landscape of Wadi Allaqi in the last few years has been the introduction of farms. Traditionally, agriculture has not been an activity undertaken by Bedouin pastoralists. However, for the reasons outlined above, in recent years women have been recognizing the merits of introducing small gardens. Women have come to believe that the production of a small area of crops can be useful in reducing the pressures created by the seasonal availability of natural grazing resources. In addition, such crops also reduce the impacts of the annual variability in the amount of grazing from one season to the next. Enthusiasm for such cultivation is more clearly shown by some Sadinab and Fashikab women, but less so by the Hamidab. Small gardens were initially set up by widows but these have now grown in popularity, spreading first to all women in Wadi Allaqi and subsequently to all households.

Those women who were initially less enthusiastic about cultivation had a range of reasons for their position. The severest difficulty that they identified was the problem of sufficient water resources for cultivation once the lake-shore started to retreat significantly in the spring. At one point, the Aswan governorate located a number of hand-pumps at the foot of both Wadi Umm Ashira and Wadi Quleib, but they proved to be unreliable: two of the three pumps in Quleib had broken within months of their installation, and others did not last much longer. Such unreliability made women wary of investing time and effort into establishing and maintaining a small farm for fear that the water supply would dry up. There was also the problem of domestic sheep and goats, as well as wild birds and animals, eating crops. Some women use dried tamarisk wood, or, more effectively if available, fish-netting as fencing material in an attempt to keep animals and birds off the crops. A further opinion, initially widely expressed, was that the amount of effort that had to go into preparing and maintaining a plot would exceed the returns it provided. The general view seemed to be that growing crops to supplement the family diet (*karkaday* [hibiscus], watermelon, tomato, and so on) would be more useful than the production of crops to supplement the sheep's and goats' diet of wild plants. The final argument, offered by relatively wealthy Hamidab women, against women producing farms was that women were not good at cultivation. However, it is not clear to what extent this opinion is linked to two further issues at the time specific to these women living in Wadi Umm Ashira. The first is that the women were of the view that the soil is salty in this location (citing the emergence of a white layer forms on top of the soil when they irrigated); and, second, an influential Habadab man had tried many times in the past to cultivate in this area, but with only limited success (perhaps because of the first factor).

The negative perception of cultivation varies across different households, and it seems possible that this may be related to individual wealth and the availability of male labor. Those women without male relatives, or whose male relatives may spend inordinate amounts of time elsewhere in the desert or Aswan, are 'trapped' in Allaqi. Hence, as they cannot take advantage of resources elsewhere in the region, they may have had few alternatives but to engage in farming.

This seems to be exemplified by the fact that the most enthusiastic initial cultivators tended to be poorer women, and especially widows. The Sadinab were generally much more eager to engage in cultivation, and some had already actively cultivated maize for sheep feed. One Sadinab widow and her sister had well-tended farms long before other Bedouin in Wadi Allaqi. They grew *karkaday*, *ful*, spices, *dura* (sorghum), *gawirma* (small melon), and watermelon, and were always enthusiastic about further possibilities for cultivation. Although widows receive LE70 per month in income support from the government, this is insufficient. Even if they have some animals to sell, and relatives who are willing to sell the animals for them, the animals' poor condition from limited quality grazing resources means that they do not command particularly high prices at market. In this situation, widows are more likely to feel that growing fodder for the animals, as well as supplements for their own diet, is worth the effort of cultivation.

Responses to women's early attempts at agriculture were not generally positive. Initial interest was limited to poorer women, and those who tended to be concentrated in the Sadinab clan. Men from households in the better-off Hamidab and Fashikab clans tend to be absent more frequently, and for longer periods, than those of the poorer Sadinab. This is primarily because the latter are much less mobile owing to their lack of camels or lack of finance to secure other forms of transport, either for transhumance or for business in Aswan. Because this is the clan whose households own few camels between them, and as a result are rarely able to concentrate the resources to take their flocks to find grazing away from Wadi Allaqi, their entire flocks are highly dependent on the limitations of the lakeshore grazing. Perhaps more than any others, these households are acutely aware of the limitations of their immediate environment, and none more so than the women of this clan. Here, where men's mobility is nearly as constrained as that of Bedouin women, their access to resources is similarly constrained to their immediate environment. In this community both men and women were more enthusiastic about the possibilities for small-scale agricultural production.

However, in the longer-term views have changed. Within three years of the first farm being established by widows, all the women had set up small gardens. These temporary agricultural areas were generally only a few meters square, and had to be carefully timed so as to fit into the seasonal variations of lake height to avoid both drought and inundation. Some women shared petrol-driven pumps to extend the growing season, and used fishing nets to prevent crop loss to animals. By 2004, these had been augmented by a large collective farm shared between Sadinab and Bishari Bedouin. This large field surrounds a water pump to take water from a well to irrigate the fields, and is used almost exclusively to grow fodder crops such as *dura* rather than plants for household consumption. These farms are still necessarily seasonal because the fluctuations in water levels of the lake even influences the water availability in the well, but the idea of agriculture is now firmly embedded as a central activity for the Bedouin community in Wadi Allaqi. Men as well as women are now involved, and the product of fodder crops is possibly now more important than the availability of ephemeral grazing for the animals kept around the households.

New Directions in Women's Activities

Given the arguments above about the relationship between gender roles, and the spaces through which men and women carry out their daily lives, it should be evident that changes in the organization of Bedouin households and production will necessarily have an impact on gender relations, roles and knowledges. Most significantly in the Wadi Allaqi communities, increasing sedentarization is clearly having an impact upon gender roles. In the desert, the partial sedentarization of Bedouin around Lake Nasser has given women increased responsibility for resource decision-making in the absence of their men folk. There is a clear sense of men and women's spaces in Wadi Allaqi. Although not confined to their homes, Bedouin women have access to a very localized environment around the household, which provides them with access to grazing around the lake, but with little scope for independent movement further afield.

Increased sedentarization with the formation of Lake Nasser has meant that there are some changes in women's roles within the household. These changes have been in response both to environmental changes and to contact with other groups. Most important to the Bedouin in the Eastern Desert has been a traditionally close relationship with Nubian women from the Nile Valley. For instance, when talking to Bedouin women about the cultural taboo on women milking sheep and goats, we were told that, although Nubian women could milk, Bedouin women were not able to—"yet." The indication of potential change is clear in this self-reflexive statement, which has been able to emerge through long-term contact with Nubian communities and experience of their characteristic gender relations. Now Bedouin households are changing in response to new opportunities. For instance, traditionally, Bedouin have refused to eat fish, claiming that was not part of the Bedouin diet. Now a number of families in Wadi Allaqi trade with commercial fisherman to buy fish and fresh vegetables to supplement their diets. The increasing number of outsiders coming to Wadi Allaqi these days to exploit the resources made more available via the asphalt road (fishermen, miners, and farmers) has meant that Bedouin society is more connected to wider concerns. Although Bedouin men have always traveled to markets in Aswan and elsewhere, and so heard news from beyond Wadi Allaqi, these visits were infrequent and generally excluded women. Now, to a certain extent, this information has come to Wadi Allaqi so is much more readily available. Women too now have a greater access to knowledge and information. Although there are still prohibitions on the extent of Bedouin women's interactions with strange men, they are growing in confidence and will no longer attempt to hide from unfamiliar visitors. As an illustration of this, it is now not uncommon for unaccompanied Bedouin women to speak to visitors requesting directions, rather than hiding from such strangers.

6
Sustaining Wadi Allaqi?

What has become abundantly clear from the discussion is the dynamic way in which the Bedouin communities in Wadi Allaqi have confronted, managed, and used to their advantage the new environmental opportunities afforded by Lake Nasser, and, at the same time, demonstrated an ecological understanding of those resources. However, there still remain questions about the sustainability of the current situation. In an earlier publication, the international interdisciplinary team working in Allaqi area came to the conclusion that:

> the most ecologically and, indeed economically sound method of encouraging development in the area is to build from 'below' within the framework of the existing Bedouin economy. This implies incorporating the views, aspirations, potentials and accumulated knowledge of the Bedouin into any development strategy for the area. The combination of the 'formal' science and 'people's' science in formulating a sustainable strategy for economic development is central to success in ensuring that such physical environments do not become unnecessarily degraded. Whether government planners and officials regard this type of approach to be a luxury they cannot afford remains to be seen.[1]

The future of Wadi Allaqi is thus caught between different views of what appropriate development might mean, between the scientific value of the MAB reserve and the resources that can be exploited, and between the state's view of how the area should be managed and the practices adopted by the Bedouin residents. Wadi Allaqi is undergoing change as it always has, and so a conclusion can only ever be a snapshot of this place. Instead, in this final chapter we offer some thoughts on what possible sustainable futures exist for Wadi Allaqi.

As we suggested in the introduction, while on one level it is difficult to question the ambitions of sustainable development, its application in practice is in no way straightforward. Immediately, the term conjures up contradictions between the notion of development and sustainability. In Wadi Allaqi there are tensions between the desires to protect the environment of the MAB Biosphere Reserve, to develop agricultural, mining, and fishing resources, and to support the continuation of Bedouin livelihoods.

An example demonstrates one line of tension. When Wadi Allaqi was declared a protected area, those Bedouin communities resident within the area were consulted about the proposals. However, the key decisions were taken within the context of an understanding of sustainable development very much in line with that of the Brundtland Commission. This was, however, very different from the understanding of the environment held by the Wadi Allaqi Bedouin. In line with accepted practice in sustainable development management, boundaries were drawn around different tracts of land in Wadi Allaqi to produce core and buffer zones that were to have different degrees of environmental protection. Within these boundaries there are particular conservation practices that should be legally observed. The idea of drawing boundaries around particular areas of Wadi Allaqi represented a very different understanding of environmental management than the more holistic view characteristic of Bedouin, for whom resources are defined in a more fluid manner. For Bedouin, conservation reflects both community needs and differing drought pressures on different vegetation resources at particular times, both on annual and significantly longer time-scales. Sustainability is a temporal practice for them, necessary at certain times of the year, in particular seasons, or during extended droughts, but less so during other periods. This cyclical practice of conservation is very different conceptually than the spatial definition characteristic of sustainable development, which constrains or excludes certain practices in defined geographic locations regardless of season or other practices.

We can see how this difference in approach to sustainable management manifests itself in practice, for example in conflicting understandings of the conservation of acacia trees by Bedouin and reserve managers. Acacia trees constitute a centrally important economic resource for the Bedouin and, as such, are very carefully managed to ensure sustainability. As we have seen, they provide a source of feed for livestock from naturally fallen leaves, shaken leaves, and fruit. They also provide an important source of wood for charcoal making; acacia is particularly valued for the quality of charcoal that can be made from it. Access to the various economic elements of acacia trees and bushes can be complex. From the same tree, one family may have claims to only naturally fallen leaves, while another may have access to those leaves that are dislodged when the plant is shaken, and a third to only the dead wood for charcoaling. For another tree, one family may have rights to all its production. The situation can be further complicated by the existence of some prohibitions against taking resources during some times in the year, whereas at other times resources can be removed without any such difficulty. This system, therefore, provides for a method of conservation of scarce resources, even though it may not necessarily meet the requirements of formal conservation practice. Bedouin conceptions of conservation are culturally and economically embedded and are managed in the interests of their community interests.

Prohibitions on removing acacia in the defined areas of environmental protection has led to further problems. In 1998, the annual level of water in the High Dam lake rose to unprecedented levels. As a result, about twelve mature acacias in Wadi Quleib, a major tributary of Wadi Allaqi, were inundated and subsequently died. In such circumstances Bedouin would traditionally use such the trees to make charcoal, owing to the fact that they would never again produce new wood. However, as the trees had grown within the conservation area, there was a prohibition against their use by humans. Consequently, in order to comply with the regulations imposed by the conservation area label, Bedouin were expected to leave the dead trees *in situ*. Unsurprisingly, Bedouin saw little logic in the formal position of conservation. There was a clear cultural divide between the two rather different views of conservation.

A potentially interesting and novel way of Bedouin using their indigenous environmental knowledge of the Wadi Allaqi Biosphere Reserve, and gaining financial reward while also supporting the preservation and conservation of this environment, is through the development of ecotourism where all stakeholders, and principally the local Bedouin community, can play an essential role. Nationally, ecotourism in the Egyptian desert has become a growing and important alternative to coastal and marine tourism, bearing in mind that 96 percent of Egypt is desert. The idea of desert tourism is not new; more than a quarter of a century ago Drouhin wrote: "desert tourism is already growing and will expand increasingly to become a flourishing industry in the future."[2] The desert attracts people by its environment, scenery, and unusual and striking rock forms. It also attracts people by its specially adapted forms of flora and fauna, and remains of dwellings of prehistoric human inhabitants, whose ruins are remarkably well-preserved in the dry desert air.[3] However, the fragility of the desert ecosystem and its slow recovery from any damage renders it extremely vulnerable to inappropriate use. Ecotourism offered by tourist companies in the form of desert safaris, involving long travel through the desert, is difficult to control and it is hard to assess its compliance with environmental standards. Such trips can unintentionally damage biodiversity by carrying tourists to places that are forbidden to visitors (for example, core areas of the Biosphere Reserve), by off-road driving (damaging or even destroying the seed bank and causing soil erosion), and by affecting the culture of local communities (for example, watching them as part of the 'wild life' or taking photographs without permission from the Bedouin). Such activities bring neither economic benefit to the conservation areas, nor to the local Bedouin, the direct benefits accruing to the tourist companies. It is already recognized by the Egyptian Tourism Development Authority that desert safaris are currently contributing to the tremendous pressures on the sensitive environment of desert regions, and there is a need for a new type of tourism development that is environmentally sound and managed in a way that promotes the conservation and protection of natural and cultural resources.[4] An alternative to unregulated desert safaris is to use national parks and other designated protected areas for managed sustainable tourism, and for local Bedouin to be centrally involved in the planning, management, and the carrying-out of day-to-day operations.

The Wadi Allaqi Biosphere Reserve has great potential for a sustainable ecotourism. The southern part of the Eastern Desert, in which Wadi Allaqi is located, remained unknown territory for visitors prior to the construction of the asphalt road in the 1990s. Only a few explorers and researchers had visited the area, limiting cultural contacts between the local Bedouin communities and outsiders. Intensive research work conducted in Wadi Allaqi and the monitoring of the ecotone on the lakeshores provide opportunities for preparing a management program for sustainable ecotourism in this fragile desert ecosystem. For example, ecotourism activities in Wadi Allaqi might include watching and photographing its wildlife. Admittedly, Wadi Allaqi does not compare to the game parks of eastern and southern Africa and effort and patience are needed to see animals, but the opportunities for seeing desert species such as Dorcas Gazelle, Barbary sheep and even the very rare Nubian Ibex are enticing.

On the other hand, bird-watching offers much more opportunity. The lakeshores in the downstream part of Wadi Allaqi host vast numbers of winter visitor birds, among which are the White Pelican, Gray Heron and

Black Stork. The wadi is an important resting point for the White Stork on its annual migration, especially in spring, when many thousands have been observed resting in Wadi Allaqi. The lakeshore is also visited in winter by numerous wildfowl, including the Teal, Pintail, and Shoveler. Tamarisk shrubs in the flood plain provide shelter, a good perch, and a feeding zone for several small passerine birds such as the Blue-Throat in wintertime and the Graceful Warbler all the year round. During summer, bird-watchers are rewarded with the arrival of a few African species, one of which is the Yellow-billed Stork. Visiting geological and archaeological sites is another special interest tourist attraction, offering traces of prehistoric sites and ancient cultures such as rock drawings, outposts, and ruined fortresses, all abundantly present throughout Wadi Allaqi.

Special interest scientific tourism can also be offered in the Wadi Allaqi Biosphere Reserve involving, for example, conservation research, ethno-botanical, geological and archaeological studies, and the opportunity to work with groups of Bedouin. Institutions throughout the world, including those located in humid areas, are dealing with research on arid lands, desertification being one of these. Situated in an extreme arid desert, Wadi Allaqi could provide the field site for multidisciplinary groups of international researchers to conduct experiments in and monitor the desert eco-system. Points of interest include the geological formation, minerals, and plant and animal species, and their adaptation to aridity. Related to this is the promotion of environmental education and the raising of awareness on the conservation of natural resources and its uses. Education and training centers in the Biosphere Reserve could be established to improve environmental education in general and knowledge of biodiversity in particular. Conducting conferences, seminars, and workshops in Wadi Allaqi could help to promote the Biosphere Reserve concept nationally and internationally by bringing experts and decision makers onto the site and into direct contact with nature, local inhabitants, and Ecotechnie-related activities.[5]

Central to all these ideas of ecotourism development is the participation of the Wadi Allaqi Bedouin at all stages, from planning through to the management and implementation of tourism activities. This will necessarily empower the community and enable Bedouin participants to influence the direction of tourism development within their local environment. In theory, ecotourism in Wadi Allaqi could generate revenue that could be used to protect and conserve the biodiversity and natural resources that draw visitors to this particular site. Crucially, it would also generate an important source of income, and hence development, that would be sustainable and consistent with the key conservation imperatives of the Wadi Allaqi Biosphere Reserve.

However, although this may well be an attractive prospect, it has to be recognized that the future for the Bedouin communities of Wadi Allaqi is not certain. While some Bedouin men are still nomadic, many are becoming less so and are spending more and more time resident in the wadi. Women, for their part, have tended to become even more sedentary than the men, and rarely leave the immediate area. Women's geographical movement patterns in the wadi are now largely related to shifts in the location of the lakeshore caused by seasonal fluctuations in the lake's water level, and not with traditional migrations with livestock to various water and grazing locations throughout the Eastern Desert. Interestingly, some of the younger women are quite enthusiastic about moving into the permanent homes constructed by the Aswan governorate in Wadi Allaqi in an attempt to 'develop' the Bedouin lifestyle and to encourage men to follow more sedentary lives.

Yet it must be recognized that Bedouin lives have constantly been changing through time by adapting to new circumstances, and opportunities and through interactions with other groups, Bedouin or otherwise. Before the High Dam was constructed they lived closely with Nubians, helping these agriculturalists harvest their crops in return for being allowed to graze their animals on the residues. Older Bedouin women still talk fondly of their Nubian friends and the agricultural practices they learned from them. More recently, Bedouin have benefited from the construction of the asphalt road to Aswan: everything from fresh food and medicines to news of family and livestock market prices has become much more accessible. The arrival of more outsiders along with the road has increased paid employment among Bedouin men, and changed aspirations. Change, therefore, is not new to the Bedouin of Wadi Allaqi.

What is new, however, is the nature of these changes. As Bedouin have become increasingly settled in Wadi Allaqi they have taken up new opportunities. Visits to Aswan are now made much more regularly, especially by men. Trucks and pick-ups are now not uncommon sights around Wadi Allaqi, and these provide transport for sheep to the markets of Aswan. Instead of having to take the sheep on foot to Aswan, which took three to four days, they can now arrive in three or four hours and with the animals in much better condition than previously. This has encouraged some Bedouin to increase sheep output, but this, in turn, has put more pressure on the grazing resources of Wadi Allaqi and surrounding areas, raising questions of long-term sustainability. In addition, this has tended to benefit the already better-off households because, in the same way that these households traditionally had the resources to take sheep to hill grazing in the winter, as explained in an earlier chapter, they retain these advantages in being able to produce more and better sheep and in gaining access to transport.

Although these are changes that the Bedouin are managing themselves, they are also subjected to external pressures and changes. The governorate, for example, is providing the communities with permanent shelters, as well as with basic health services. However, the housing has been built at a location several kilometers from the lakeshore in an area where there are difficulties in water supply and virtually no grazing for the livestock. It remains to be seen whether the Bedouin choose to remain there or migrate back to the lakeshore.

A major threat to the long-term sustainability of Wadi Allaqi, however, is presented by some of the incomers to the area, including itinerant fishermen, quarry-workers, and the military. Whereas the Bedouin are well aware of the vulnerability and fragility of the natural resource endowment of the region, and act accordingly, many of these incomers are less circumspect. There are real worries that serious damage is steadily being done by uprooting trees for firewood and polluting water in and around the lake. The existence of the Wadi Allaqi Biosphere Reserve does at least provide some legal protection, but there must remain concerns that this may not be enough.

However, in all this story, what is clear is the resilience of the Bedouin. They have demonstrated a willingness and dynamism to take advantage of the new resource opportunities, and their economy and society have adapted accordingly. It is also clear that they retain an understanding and empathy with their surrounding environment and manage it in a broadly sustainable manner. This should be no surprise. The Bedouin live in a tough and unforgiving environment that must be carefully and sustainably managed, otherwise Bedouin society will not survive. Conservation, sustainability, and management are central to Bedouin society, indeed deeply embedded within daily activity and practice, even if the Bedouin themselves may not recognize these concepts as such. Perhaps we should acknowledge that, of all the stakeholders involved in Wadi Allaqi, perhaps it is the Bedouin who have the most appropriate sustainable economic methods, and, as the key stakeholders, should be left and encouraged to deliver the most appropriate sustainable development for the area.

Appendix 1

List of Project Researchers

Ekramy El-Abassiry
Professor Ramadan Abdallah (deceased)
Professor Abd El-Razik Abd El-Alim
Professor Said Abo El-Ella
Dr. Magdi Ali
Amal Awadallah
Dr. Samia Abd El-Aziz
Douglas Ball (deceased)
Professor Ahmed Belal
Professor Prosanto K. Biswas
Professor Reinhard Bornkamm
Professor W. Breckle
Professor John Briggs
Renee Brunelle
Professor Fathalla El-Cheikh
Dr. Gordon Dickinson
Dr. Ludmila Emelianova
Professor Nasr El-Emary (deceased)
Caryll Faraldi
Dr. David Forrest
Professor Mohamed Gabr
Professor Samir Ghabbour
Professor Hassanin Gomaa
Professor Arafa Hamed
Nabila Hamed
Dr. Jim Hansom

Professor Lotfy M. Hassan
Dr. Abdel El Salam Ibrahim
Haythem Ibrahim
Professor Naieem El-Kaltaway
Professor Arafat Kamel
Professor Mohamed Kassas
Professor Imam Khalifa
Hanaa Kondol
Dr. Brenda Leith
Professor Andrew Long
Munir Mahgoub
Professor Mohamed Raafat Mahmoud
Abdel-Moneim Mekki
Hatem Mekki
Professor Wafai Mikhail
Dr. Sayed Nour El-Din Moalla
Dr. Usama Mohalel
Awadalla Mohamed
Dr. Alastair Morrison
Dr. Kevin Murphy
Dr. Nina Novikova
Dr. Ahmed El-Otify
Dr. Ian Pulford
Dr. Magdy Abd El-Radi
Tarek Radwan
Dr. Usama Radwan

Professor Hussein Raghib
Professor Samir Riad
Dr. Alan Roe
Kassem Said
Dr. Ramadan Salem
Dr. Sayed Abdou Selim
Dr. Joanne Sharp
Dr. Abd El-Samie Shaheen
Peter Snelson
Dr. Hassan M. Sobhy
Dr. Hassan Sogheir
Dr. Mohamed Sogheir (deceased)
Professor Jacqueline Solway

Dr. Wafaa Sorour
Professor Irina Springuel
Mustafa Taher
Professor Ragaie El-Tahlawy
Dr. Hussein Tahtawy
Tiffany White
Dr. Maik Veste
Machiel de Vries
Dr. Hoda Yacoub
Dr. Magdi Younis
Eric van Zanten
Dr. Tarek Zedan

Appendix 2

List of Allaqi Project Working Papers

The Allaqi Project Working Papers series was established in the early stages of the research to provide an easily accessible repository for our results. Many of the papers presented only interim findings and were there to stimulate discussions and debate within the group, and many provided the initial basis from which research papers were subsequently produced for publication in peer-reviewed international research journals.

UESD Working Papers

1. "Water resources and eco-development," Gordon Dickinson (1989).
2. "The Soils of Wadi Allaqi: A General Overview," Ian D. Pulford (1989).
3. "Human activity in Wadi Allaqi, April, 1989—a preliminary report," John Briggs (1989).
4. "Plant ecology of Wadi Allaqi and Lake Nasser No. 1: Aquatic plants of Lake Nasser and associated waters in Egyptian Nubia," Irina Springuel, and Kevin Murphy (1989).
5. "Plant ecology of Wadi Allaqi and Lake Nasser No. 2: Preliminary vegetation survey of the downstream part of Wadi Allaqi," Irina Springuel, M.M. Ali, and Kevin Murphy (1989).
6. "Applications of remote sensing and geographical information systems in the Wadi Allaqi area, south-east Egypt," Sayed A. Ahmed Selim (1990).
7. "Interim report on soil fauna and composting in the Wadi Allaqi Project," S. Ghabbour and W. Mikhail (1990).
8. "Properties of recently inundated soils in Wadi Allaqi," U.A. Radwan, A.S. Shaheen, and I. Pulford (1990).
9. "Geology, mineral resources and water conditions in the Wadi Allaqi area," Abd al-Moneim Mekki and Gordon Dickinson (1990).
10. "Plant ecology of Wadi Allaqi and Lake Nasser No. 3: Flora of the Wadi Allaqi Basin," I. Springuel, L.M. Hassan, M. Sheded, M. al-Soghir Badri, and M.M. Ali (1991).

11. "The economic system of Wadi Allaqi," Abd al-Moneim Mekki and John Briggs (1991).
12. "The Birds of Wadi Allaqi—Lake Nasser, Egypt," Samia Abd al-Azeiz and John Walmsley (1991).
13. "Plant ecology of Wadi Allaqi and Lake Nasser No. 4: Basis for economic utilization and conservation of vegetation in Wadi Allaqi conservation area, Egypt," Irina Springuel (1991), (second edition, 1994).
14. "Provisional atlas of Allaqi," ed. John Briggs (1991).
15. "Conservation management strategies for the Wadi Allaqi Protected Area," Gordon Dickinson (1991).
16. "The Social and demographic structure of Wadi Allaqi," A.I. Mohamed, A.M. Mekki and J. Briggs (1991).
17. "The Allaqi project management and output," Ahmed Esmat Belal (1992).
18. "Survey of soil resources in Wadi Allaqi," Sayed M.N. Moalla and Ian D. Pulford (1993).
19. "Processes influencing soil formation and properties in Wadi Allaqi," S.M.N. Moalla and I. Pulford (1993).
20. "Economic value of desert plants acacia trees in Wadi Allaqi conservation area," Irina Springuel and Abdel-Moneim Mekki (1993).
21. "Medicinal plants in Wadi Allaqi," Irina Springuel, Nasr al-Emary, and A.I. Hamed (1993).
22. "Studies of soil fauna in Wadi Allaqi," S.I. Ghabbour, W.Z.A. Mikhail, and H.M. Sobhy (1993).
23. "The plant ecology of Lake Nasser and Wadi Allaqi: Factors influencing the establishment and survival of acacia seedlings in Wadi Allaqi," Abd al-Samia Shaheen, I. Springuel, and K.J. Murphy (1993).
24. "Fuelwood resources in Wadi Allaqi: Social, economic, and ecological aspects," Jacqueline S. Solway (1995).
25. "Overgrazing in Wadi Allaqi," Tiffany White (1995).
26. "Biomass determination and growth model of *Tamarix nilotica* shrub in Wadi Allaqi, Lake Nasser area, Aswan, Egypt," H. Marei, R. Salem, A. Long, and A. Belal (1995).
27. "Vegetation types of an extensive drainage system in the South-Eastern Desert, Egypt: A multivariate analysis," M.M. Ali, M. Badri, L. Hassan, and I. Springuel (1995).
28. "Environment and producers in Wadi Allaqi ecosystems," Irina V. Springuel (1995).
29. "Physical and chemical characteristics of fuel wood species in Wadi Allaqi, Lake Nasser area, Egypt," A. Belal, I. Pulford, H. Marei, S.D. McGregor, H. El-Shaikh and R. Salem (1995).
30. "Vegetation uses and ecological values in Wadi Allaqi: A participatory rural appraisal approach," John Briggs and Mohamed Badri (1997).

31. "Soil salinity and requirements for irrigation in Wadi Allaqi," U. Radwan, T. Radwan, M. Sheded, and I. Pulford (1997).
32. "Indigenous and scientific knowledges: The choice and management of cultivation sites by Bedouin in Upper Egypt," John Briggs, Ian Pulford, Mohamed Badri, and Abd al-Samia Shaheen (1998)
33. "Socio-economic system of Wadi Allaqi," Jacqueline Solway and Abd al-Moneim Mekki (1999).
34. "Sheep production in Wadi Allaqi," Hatem Abd al-Monaim, Abd al-Monaim Mekki, and John Briggs (2000).
35. "Cultivation of *Balanites* and *Acacia* in the Allaqi farm," Irina Springuel, Mohamed Badri, Usama Radwan, Tarek Radwan, Hatem Mekki, and Mostafa Taher (2000).
36. "Palatability of desert fodder plants," I. Springuel, H.A. Hussein, M. al-Ashri, M. Badri, and A. Hamed (2001).
37. "Women, environmental knowledge and livestock production in Wadi Allaqi biosphere reserve," I. Springuel, J. Briggs, J. Sharp, H. Yacoub, N. Hamed (2001).
38. "Environmental management of African arid lands (Special publication)," ed. G. Dickinson and I. Pulford (1990?).

Appendix 3

List of Allaqi Project Publications

Abd al-Wahab, H., A. Hamed, and N. al-Emary. 1998. Antitermite principles isolated from the wild herb, *Psoralea plicata* Del. *Assiut University Bulletin for Environmental Researches* 1 (2): 17–25.

Ali, M.M., M.A. Badri, L.M. Hassan, and I. Springuel. 1997. Effect of physiogeographical factors on desert vegetation, Wadi Allaqi Biosphere Reserve, Egypt: A multivariate analysis. *Ecologie* 28 (2): 119–28.

Ali, M.M., M.A. Badri, S.N. Moalla, and I.D. Pulford. 2001. Cycling of metals in desert soils: Effects of *Tamarix nilotica* and inundation by lake water. *Environmental Biochemistry and Health* 23 (4): 369–78.

Ali, M.M., G. Dickinson, and K.J. Murphy. 2000. Predictors of plant diversity in a hyperarid desert wadi ecosystem. *Journal of Arid Environments* 45:215–30.

Ali, M.M., A.M. Hamad, I. Springuel, and K. Murphy. 1995. Environmental factors affecting submerged macrophyte communities in regulated water bodies in Egypt. *Archiv für Hydrobiologie* 133 (1): 107–28.

Badri, M. and A. Hamed. 2000. Nutrient value of some plants in an extremely arid environment (Wadi Allaqi Biosphere Reserve, Egypt). *Journal of Arid Environment* 44:347–56.

Badri, M.A., I.D. Pulford, and I. Springuel. 1996. Supply and accumulation of metals in two Egyptian desert plant species growing on wadi-fill deposits. *Journal of Arid Environments* 32:421–29.

Badri, M. and I. Springuel. 1994. Biogeochemical prospecting in the South Eastern Desert of Egypt. *Journal of Arid Environments* 28:257–64.

Belal, A. 1993. Sustainable development in Wadi Allaqi. *Desertification Control Bulletin* 23:39–43.

Belal, A.E., I.D. Pulford, H. Marei, S.D. McGegor, H.A. Elshaika, and R. Salem. 2000. Physical and chemical characteristics of fuel wood species in Wadi Allaqi, Lake Nasser area, Aswan, Egypt. *Science and Technology Bulletin* 21.

Belal, A.E. and I. Springuel. 1996. Economic value of plant diversity in arid environments. *Nature and Resources* 32 (1): 33–39.

Belal, A. and I. Springuel. 1997. *Wadi Allaqi Biosphere Reserve* (booklet). Cairo: UNESCO.

Briggs, J. 1991. Pastoralism in Wadi Allaqi region, Egypt: An economy under pressure. In *Pastoral economies in Africa and long term responses to drought*, ed., J. Stone, 197–205. Aberdeen: Aberdeen University Press.

Briggs, J. 1994. Development in the desert: The Wadi Allaqi, Egypt. In *The Arab world*, ed., A. Findlay, 138–41. London: Routledge.

Briggs, J. 1995. Environmental resources: Their use and management by the Bedouin of the Nubian Desert of southern Egypt. In *People and environment in Africa*, ed., J.A. Binns, 61–67. Chichester, UK: John Wiley.

Briggs, J., M. Badri, and A.M. Mekki. 1999. Indigenous knowledges and vegetation use among Bedouin in the Eastern Desert of Egypt. *Applied Geography* 19:87–103.

Briggs, J., G. Dickinson, K. Murphy, I. Pulford, A.E. Belal, S. Moalla, I. Springuel, S.I. Ghabbour, and A.M. Mekki. 1993. Sustainable development and resource management in marginal environments: Natural resources and their use in the Wadi Allaqi region of Egypt. *Applied Geography* 13:259–84.

Briggs, J., I. Pulford, M. Badri, and A.S. Shaheen. 1998. Indigenous and scientific knowledges: The choice and management of cultivation sites by Bedouin in Upper Egypt. *Soil Use and Management* 14:240–45.

Briggs, J., J. Sharp, N. Hamed, and H. Yacoub. 2003. Changing gender relations, changing environmental knowledges and livestock management in Upper Egypt. *Geographical Journal* 169:313–25.

Briggs, J., Sharp, J., Yacoub, H., Hamed, N. and Roe, A. 2007. The nature of indigenous knowledge production: evidence from Bedouin communities in southern Egypt. *Journal of International Development*, 19, 239–251.

Dickinson, G., K. Murphy, and I. Springuel. 1993. The implications of the altered water regime for the ecology and sustainable development of Wadi Allaqi, Egypt. In *Environmental changes in drylands: Biogeographical and geomorphological perspectives*, eds. A.C. Millington and K. Pyke, 379–91. Hoboken, NJ: John Wiley and Sons Ltd.

Hamed, A., S. Piacente, G. Autore, S. Marzocco, C. Pizza, and W. Wieslaw Oleszek. 2005. Antiproliferative hopane and oleanane glycosides from roots of *Glinus lotoides* L. *Planta Medica* 71 (6): 554–60.

Hamed, A.I., I. Springuel, and N.A. El-Emary. 1998. Benzofuran glycosides from *Psoralea plicata*. *Phytochemistry* 50:887–90.

Hamed, A., I. Springuel, N.A. El-Emary, H. Mitome, H. Miyaoka, and Y.A. Yamada. 1996. Triterpenoidal Saponin Glycosides from *Glinus lotoides* var. *dictamnoides*. *Phytochemistry* 43 (1): 183–88.

Hamed, A., I. Springuel, N. El-Emary, H. Mitome, H. Miyaoka, and Y.A. Yamada. 1997. Phenolic cinnamate dimer from *Psoralea plicata* Del. *Phytochemistry* 45:1257–61.

Khalil, A., S. Hammad, A. Belal, and A. Mohamed. 1995. Physical-chemical and mineralogical characterization of High Dam Lake sediments. *Tile and Brick* 11 (6): 446–49.

Mikhail, W.Z. 1993. Effect of soil structure on soil fauna in a desert wadi in southern Egypt. *Journal of Arid Environment* 24:321–31.

Moalla, S.N. and I.D. Pulford. 1995. Mobility of metals in Egyptian desert soils subject to inundation by Lake Nasser. *Soil Use and Management* 11:94–98.

Pulford, I., K. Murphy, G. Dickinson, J. Briggs, and I. Springuel. 1992. Ecological resources for conservation and development in Wadi Allaqi, Egypt. *Botanical Journal of the Linnean Society* 108:131–41.

Radwan, U.A. 2006. Effects of NaCl and $CaCl^2$ treatment on stomatal conductance and leaf water potential of *Solenostemma arghel* Del. Hayne under different levels of photosynthetic photon flux density (PPFD). *Journal Union Arab Biologist* 17:1–11.

Radwan, U.A., I. Springuel, and P.K. Biswas. 2005. The effect of salinity on water use efficiency of *Solenostemma arghel* (Del.) Hayne. *Journal Union Arab Biology* 16B:129–45.

Radwan, U.A., I. Springuel, P.K. Biswas, and G. Huluka. 2001. The effect of salinity on water use efficiency of *Balanites aegyptiaca*. *Egyptian Journal of Biology* 2:1–7.

Shaddad, M.A.K., I. Springuel, S-W. Breckle, and W.A. Sorour. 2001. Drought and salinity tolerance of desert and riverine population of *Faidherbia albida*. *Bulletin Faculty of Science, Assiut University* 30(2–D): 245–60.

Sharp, J., J. Briggs, H. Yacoub, and N. Hamed. 2003. Doing gender and development: GAD, empowerment and local gender relations. *Transactions, Institute of British Geographers* 28 (3): 281–95.

Sheded, M.G., I. Pulford, and A.I. Hamed. 2006. Presence of major and trace elements in seven medicinal plants growing in the South-Eastern Desert, Egypt. *Journal of Arid Environment* 66:210–17.

Springuel, I. 1991. Bases for the economic utilisation and conservation of vegetation in Wadi Allaqi conservation area, Egypt. *African Journal of Agricultural Science* 18 (2): 65–84.

Springuel, I., 1993. Riparian vegetation in an hyper-arid area in Upper Egypt, Lake Nasser area, and its sustainable development." *Proceedings of the international workshop on the ecology and management of aquatic-terrestrial ecotones*. University of Washington, Seattle, U.S.

Springuel, I. 1996. Environment and producers in Wadi Allaqi ecosystem. *Arid Ecosystems* 2–3:43–57.

Springuel, I. 1997. Vegetation, land use and conservation in the Southern Eastern Desert of Egypt. In *Reviews in ecology: Desert conservation and development*, eds., H. Barakat and A. Hegazy, 177–206. Cairo: Metropole.

Springuel, I. 2001. Indigenous agroforestry for sustainable development of the area around Lake Nasser, Egypt. In *Sustainable land use in deserts*, eds., S.W. Breckle, M. Veste, W. Wucherer. Berlin: Springer.

Springuel, I. and M.M. Ali. 1991. Impact of Lake Nasser on desert vegetation. In *Desert development. Part 1: Desert agriculture, ecology and biology. Proceedings of the Second International Desert Development Conference held on 25–31 January 1987 in Cairo Egypt*, eds., Adly Bishay and Harold Dregne, 557–68. New York: Harwood Academic Publishers.

Springuel, I. and A. Belal. 2002. Ecotourism in the Wadi Allaqi Biosphere Reserve. *Biosphere Reserves Bulletin* 11:26–27. See the full text at www.unesco.org/mab/qualityEconomies/WadiAllaqi.pdf.

Springuel, I. and A. Belal. 2003. Wadi Allaqi biosphere reserve. *Prospects* 33 (3): 325–37.

Springuel, I., N. al-Hadidi, and M. Sheded. 1991. Plant communities in the southern part of Eastern Desert (Arabian Desert) of Egypt. *Journal of Arid Environments* 21:307–17.

Springuel, I., A.H. Hussein, M. El-Ashri, M. Badri, A. Hamed. 2003. Palatability of fodder plants. *Arid Ecosystems* 9 (18): 30–39.

Springuel, I. and A.M. Mekki. 1994. Economic value of desert plants: acacia trees in the Wadi Allaqi Biosphere Reserve. *Environmental Conservation* 21 (1): 41–48.

Springuel, I. and N. Novikova. 2006. Ecotonal vegetation (aquatic-terrestrial) in upper part of river Nile. *Arid Ecosystems* 12 (30): 39–48.

Springuel, I., U.A. Radwan, P.K. Biswas, D. Hileman, G. Huluka, and M. Alemayehu. 1992. Water conditions of the soils in Wadi Allaqi, Lake Nasser region. In *Proceedings of the first national conference on land reclamation and development in Egypt*, ed., Mohamed Atef Kishk, 307–26. Minia, Egypt.

Springuel, I., A.S. Shaheen, and K.J. Murphy. 1995. Effects of grazing, water supply, and other environmental factors on natural regeneration of *Acacia raddiana* in an Egyptian Desert Wadi System. In *Rangelands in a sustainable biosphere: proceedings of the fifth International Rangeland Congress* 1:529–30, ed. Neil E. West. Colorado: Society for Range Management.

Springuel, I. and M. Sheded. 1990. Spatial analysis of the plant communities in the southern part of the Eastern Desert, Egypt. *Journal of Arid Environments* 21:319–25.

Springuel, I., M. Sheded, and K.J. Murphy. 1997. The plant biodiversity of the Wadi Allaqi Biosphere Reserve (Egypt): Impact of Lake Nasser on desert wadi ecosystem. *Biodiversity and Conservation* 6:11.

Notes

Notes to Introduction

1   Al-Khodari 2003.
2   Kassas 2002.
3   White 1988.
4   Dafalla 1975.
5   Al-Sayed and van Dijken 1995.
6   Environmental education is the logistic function of the Wadi Allaqi
    Biosphere Reserve (WABR) implemented by the Unit of Environmental
    Studies and Development (UESD). It has strengthened the building
    capacity of WABR by creating a strong team of multidisciplinary
    researchers who are capable of working in a desert environment.

Notes to Chapter 1

1   Wadi Allaqi was declared a conservation area in 1989 by Prime Minister's
    Decree under Law No. 102/1983 and has had protected status since
    then within the Egyptian Environmental Affairs Agency. WAPA is in a
    national protected area network with the main goals of "Maintaining
    the diversity and viability of the various components of Egypt's natural
    heritage, to ensure their sustainable utilization, through conserving
    adequate representative examples of the country's natural ecosystems
    and landscapes, for the benefit of present and future generations: the
    intergenerational equity" (A Status Report on the Protected Area Network
    of Egypt, 2003).
2   Hoath 2003.
3   Passage of Law 102 concerning protected areas in 1983.
4   Bond 1989.
5   Baha al-Din, Sherif M. 1999, Directory of Important Bird Areas in Egypt,
    Birdlife Int., The Palm Press, Cairo.
6   Wadi Gabgaba, joins the main trunk of Allaqi in its downstream stretch.
    This wadi extends southward into the extremely dry desert of North
    Sudan, and was probably a channel of the older phase of the River Nile.

7   UNESCO MAB, International co-ordinating Council of the Program on the Man and the Biosphere, Final Report. Paris, 9–17 November 1971.

8   Seville Strategy 1995.

9   http://www.unesco.org/mab/mabProg.shtml

10   News from the Biosphere World Network, No. 2, January 2007, http://www.unesco.org/mab/publications/newsletter/eng.shtml

11   Haithem Ibrahim, personal unpublished record.

12   Khalil and Ali 2003.

13   We can refer to Findlay (1998): "strategies to conserve arid environments for future generations, such as the establishment of park and areas where grazing of livestock is prohibited, while not totally inappropriate, may have very adverse effects for current pastoral populations by denying them from traditional grazing rights."

14   Under the national land-reform policies in the 1950s and 1960s, the desert regions were declared state-owned and open to unrestricted use by all citizens. Thus, the pastoral tribes lost their rights to control and manage the rangelands and associated water supplies. This demonstrates the process by which tribal authority has been usurped by the state, traditional social control mechanisms weakened, and pastoral communities marginalized.

15   (1) Conservation: the Nature Conservation Sector (NCS) under the EEAA is the government body responsible for nature conservation and management of protected areas. (2) Developers: the state (the Aswan Lake Nasser Development Authority [ALNDA] and, to some extent, the Ministry of Agriculture and the Aswan governorate) and private capital investment (fishermen, investors, workers and technical staff of mines and quarries). (3) Research Institutions: UESD has conducted research, indigenous knowledge studies, monitoring, pilot demonstration projects, training, informal education through excursions, and students' practical work (researchers, educators, students, technical staff). (4) The local community, comprising Bedouin who are living in Wadi Allaqi and have an interest in the utilization of natural resources.

16   The Ministry of Defense controls the entire border with Sudan and is responsible for combating illegal smuggling across the entire border. Army camps and barracks are spread in different areas throughout the entire region of the Biosphere Reserve.

17   Australia's Gippsland Ltd. said it had signed a contract with Perth-based contractors Wallis Drilling to complete up to 14,000 meters of drilling in the possibly lucrative gold and copper-nickel prospects in the Wadi Allaqi region located southeast of Aswan, Egypt. The three-month program will test eleven historical gold mines and workings, plus some targets identified by recent geochemical exploration.

18   Healy 1997.

19   Springuel and Mekki 1994.

Notes to Chapter 2

1   Raheja 1973.
2   White 1988.
3   White 1988.
4   Springuel and Ali 2005.
5   Said 1993.
6   Latif 1974.
7   Comparing the results of microbiological surveys of the lake carried out in 1974, 1978 and 1996, there had been a remarkable increase in the total bacterial counts during the last two decades. This had amounted to more than a thousandfold with particularly high values having been recorded near the dam (Bishai, 'Abd al-Malek, and Khalid 2000).
8   The average catch of fish in the reservoir increased from 2,600 metric tons in 1968 to 22,500 metric tons in 1978 to 34,000 metric tons in 1987. However, the fish catch dramatically decreased in the early 1990s but it was unclear whether this was the result of decreasing fertility of the system or overexploitation. "Lake Nasser was providing adequate supplies until the early 1980s, when production started to plummet. Over the last two decades fishermen have proceeded with their work despite the steady decrease in the quantity of fish they produce—from 34,000 tonnes in 1981 to a mere 4,000 in 2000," *Al-Ahram Weekly*, 31 March–6 April 2005, issue no. 736.
9   Bishai, 'Abd al-Malek, and Khalid 2000.
10 Kassas 2002.
11 Bishai, 'Abd al-Malek, and Khalid 2000.
12 Soil moisture was monitored by using the neutron probe. The research on the vertical water movement in soil was conducted in the Wadi Allaqi in the period from 1992 to 1995 until all of the flood plain was covered by rising lake water in 1996. Results of this study were presented in Springuel, Radwan, Biswas, Hileman, Huluka, and Alemayenu 1992; "Cultivation of Medicinal Plants" annual report (1995).
13 The effect of the lake on hydrology of Wadi Allaqi was part of the M.Sc. research of Machil de Vries and Eric van Zanten, postgraduate students form Institute of Earth Sciences, Free University, Amsterdam. They did three months field studies in Wadi Allaqi in period from November 1991 until January 1992 and published the results of research in the report "A hydrogeological survey of the Wadi Allaqi, Egypt with emphasis on the influence of the Aswan High Dam Lake," 1993, Institute of Earth Science, Free University, Amsterdam. This report is the most reliable source of information on the groundwater resources in the downstream part of the Wadi Allaqi to which we refer in this chapter together with information obtained by Allaqi research team in different periods.
14 Springuel, Murphy, and Sheded 1997; Ali, Badri, Hassan, and Springuel 1997.

15 Noy-Meir 1973.
16 Piotrovsky 1965.
17 Muhammad Ahmad ibn Abd Allah, a Nubian born near Dongola, organized an Islamic popular movement that culminated in the founding of the Mahdist state that dominated present-day Sudan from 1885 to 1898.
18 Mekki and Dickinson 1990.
19 De Vries and van Zanten 1993.
20 De Vries and van Zanten 1993.
21 Installation of piezometers was supported by UNEP, project no: 5201-88-81-2223.
22 "Cultivation of Medicinal Plants" mid-year report (1995).
23 Our studies showed that water remained in the wadi deposits on the depth 8–10 meters for more than 10 years.
24 De Vries and van Zanten 1993.
25 Mekki and Dickinson 1990.
26 De Vries and van Zanten 1993.
27 EC is the measure of the total concentration of the dissolved salts in the water. When salts dissolve in the water, they give off electrically charged ions that conduct electricity. The more ions in the water, the greater electrical conductivity it has. The fresh water has low EC, below 0.2 dS cm^{-1}, in comparison with ocean water with 42 dS cm^{-1}.
28 Kapetsky J.M. and Peter T., eds. *Status of African reservoir fisheries*. Fishery Resources and Environment Division, Fisheries, Department, FAO Rome 1984. http://ww.fao.org/docrep/008/ad795b/AD795B15.htm.
29 De Vries and van Zanten 1993; Mekki and Dickinson 1990.
30 References used: Pulford 1989; Radwan, Shaheen and Pulford 1990; Moalla and Pulford 1993a;
Moalla and Pulford 1993b; "Cultivation of Medicinal Plants" mid-year report (1995).
31 De Vries and van Zanten 1993.
32 Moalla and Pulford 1993b.
33 De Vries and van Zanten 1993.
34 Mekki and Dickinson 1990.
35 Briggs, Dickinson, Murphy, Pulford, Belal, Moalla, Springuel, Ghabbour and Mekki 1993.
36 Pulford, Murphy, Dickinson, Briggs, and Springuel 1992.
37 Moalla and Pulford 1993b.
38 Pulford, Murphy, Dickinson, Briggs, and Springuel 1992.
39 Radwan, Shaheen, and Pulford 1990.
40 Moalla and Pulford 1995.
41 Radwan, Shaheen and Pulford 1990.
42 Pulford, Murphy, Dickinson, Briggs, and Springuel 1992.
43 Moalla and Pulford, 1993.

44 Pulford, Murphy, Dickinson, Briggs, and Springuel 1992.
45 Pulford, Murphy, Dickinson, Briggs, and Springuel 1992.; Briggs, Dickinson, Murphy, Pulford, Belal, Moalla, Springuel, Ghabbour and Mekki 1993.
46 Moalla and Pulford 1993a.
47 142 trees (67 trees of *Acacia raddiana*, 16 trees of *Acacia tortilis* and 28 *Balanites aegyptiaca* trees and 31 *Salvadora persica* shrubs) were counted in the main channel of wadi on the total area of 200,000 m² (20 ha).
48 Springuel 1997.
49 Briggs, Dickinson, Murphy, Pulford, Belal, Moalla, Springuel, Ghabbour and Mekki 1993.
50 Nykänen-Kurki, P., P. Leinonen and A. Nykänen. Dry matter production of annually cultivated forage legumes in organic farming. http:/orgprints. ord/4293/01/4293doc.
51 Quoting from Kassas (1952): "This type of vegetation may be called accidental vegetation in distinction from the ephemeral annual vegetation and the perennial or permanent vegetation that are met within deserts with a more regular rainfall." In the South Eastern Desert where rain occurs once in a few years, only accidental vegetation can exist in the habitats which have no access to the groundwater. These species are potential annuals or potential perennials which are able to perform an annual life-cycle but can likewise continue growing as long as water persists in the soil.
52 Marei, Salem, Long, and Belal 1995.
53 Springuel, Shaheen, and Murphy 1995.

Notes to Chapter 3

1   Hall 1992.
2   The research on the fuelwood resources in Wadi Allaqi was conducted from 1992 to 1998 in co-operation with Trent University, Canada. Two projects were supported by International Development Research Center (IDRC) Canada. Environmental management of fuelwood resources in Wadi Allaqi, Egypr, The final report file 921001-01, 1995 and Environmental valuation and management of plants in Wadi Allaqi, Egypt, the final report file 95-1005-01/02 127-01 1998 are the main source of information used in this chapter.
3   Studies conducted in Wadi Allaqi showed that Bedouns most valued tamarisk for providing shade both for small stock and for human settlement (Briggs, Badri, and Mekki 1999).
4   Briggs and Badri 1997.
5   Belal, Pulford, Marei, McGregor, El-Shaikh, and Salem 1995.
6   Part on the fodder plants summarized information from following publications: Irina Springuel 1991; Belal, Leith, Solway, and Springuel, 1998; Springuel, Hussein, al-Ashri, Badri, and Hamed 2003.

7   Hamed, Ibrahim, Mekki, Springuel, Yacoub, Briggs, Roe, and Sharp 2002.

8   Krzywinski, Knut. 2001. 'Culture Landscapes" in K. Krzywinski and
    R.H. Pierce, eds., Deserting the desert. Norway: Alvheim and Eide
    AkademiskForlag.

9   A.H. Hussein (1997), who studied the properties of fodder plants in the
    Wadi Allaqi flood plain found that in December 1994 (after the rain of
    November 1994), the protein content of forage (crude protein [CP] about
    10 percent in *Acacia raddiana* and 12 percent in *Acacia ehrenbergiana*) was
    higher than in the dry December of 1995 (6 percent and 5 percent for
    A.raddiana and *A. ehrenbergiana* respectively) when the rains failed for the
    year. Accordingly, the dry matter digestibility and organic matter digestibility
    (DMD and OMD respectively) decreased in dry season (December 1995) in
    both *Acacia* species, and was especially noticeable for *A. raddiana*.

10 Springuel 1994.

11 Table not given, for reference, see Hussein 1997.

12 For chemical composition of *Glinus* and *Hyoscyamus muticus*, see Hussein
    1997.

13 Hussein 1997.

14 Tannins precipitate proteins by hydrogen bonding and hydrophobic
    interactions to form stable complexes at rumen pH, which adversely affect
    protein and fiber digestion in the rumen and thereby protein availability
    to the animal (Ramana, Singh, Solanki, and Negi 2000).

15 Kumar R. and D'Mello J.P.F. 1995; Degen, Mishorr, Makkar, Kam,
    Benjamin, Becker, and Schwartz 1998.

16 Hamed, Springuel, El-Emary, Mitome, Miyaoka, and Yamada 1996;
    Hamed and El-Emary 1999.

17 Adam, al-Yahya, and al-Farhan 2001.

18 Moalla and Pulford 1995.

19 Pulford, Murphy, Dickinson, Briggs, and Springuel 1992.

20 The soil studies were conducted in order to select the site for
    establishment farm for cultivation of medicinal plants during the
    implementation of project on studies of medicinal plants in Wadi Allaqi.
    Results of these studies are published in the mid-year report "Cultivation
    of Medicinal Plants" submitted to the Ministry of International Co-
    operation, Assiut, Egypt 1995.

21 "Cultivation of Medicinal Plants" mid-year report (1995).

22 Radwan, Shaheen, and Pulford 1990.

23 Radwan, Radwan, Sheded, and Pulford 1997.

24 Radwan, Radwan, Sheded, and Pulford 1997.

25 Available water and rich biomass attract numerous pests to the cultivated
    fields from the surrounding desert. The biological control, which is
    alternative to chemical pesticide, needs the large investment including
    research work.

26 Springuel 2001.

Notes to Chapter 4

1   Material in this chapter is based on Briggs 1989; Mohamed, Mekki, and Briggs 1991; Briggs 1991; Murphy, Pulford, Dickinson, Briggs and Springuel 1992; Briggs, Dickinson, Murphy, Pulford, Belal, Moalla, Springuel, Ghabbour and Mekki 1993; Briggs 1995; Briggs and Badri 1997; Briggs, Badri, Pulford, and Abdel-Samir 1998; Briggs, Badri, and Mekki 1999; Abdel-Monaim, Mekki, and Briggs 2000; Springuel, Briggs, Sharp, Yacoub, and Hamed 2001; Briggs and Sharp 2004; White 1995; Briggs, Sharp, Yacoub, Hamed, and Roe 2007.

2   A lambing percentage of 100 implies that 100 breeding ewes produce 100 lambs in a season. A lambing percentage of 200 means that the equivalent 100 breeding ewes produce 200 lambs and therefore have many multiple births. A lambing percentage of 50 shows that production is poor with only 50 lambs being produced by 100 breeding ewes.

Notes to Chapter 5

1   Material in this chapter is based on Springuel, Briggs, Sharp, Yacoub, and Hamed 2001; Sharp, Briggs, Yacoub, and Hamed 2003; Briggs, Sharp, Hamed, and Yacoub 2003; Briggs and Sharp 2004; Solway and Mekki 1999.

Notes to Chapter 6

1   Briggs, Dickinson, Murphy, Pulford, Belal, Moalla, Springuel, Ghabbour, and Mekki 1993.

2   Drouhin 1970.

3   Sutton 1976.

4   Tourism Development Authority 1999.

5   Captain Jacques Cousteau proposed the Ecotechnie concept, which aims holistically to consider ecology, economics, the social sciences and technology in order to understand the long-term consequences of management and development decisions. Ecotechnie is a term that comprises existing interdisciplinary efforts in the field of environment and development, including, but not limited to, ecological economics, human ecology and eco-technology. In 1994, Captain Cousteau initiated the UNESCO-Cousteau Ecotechnie Program, focusing on supporting and enhancing interdisciplinary education in the environmental field.

References

Abdel-Monaim, H., A.M. Mekki, and J. Briggs. 2000. Sheep production in Wadi Allaqi. Allaqi Project Working Paper Series No. 34.

Abu-Al-Futuh, I.M. 1983. *Balanites aegyptiaca*, An unutilized raw material potential ready for agro-industrial exploitation. UNIDO Report, TF/INT/77/021, distribution limited.

Adam, S.I.A., M.A. Al-Yahya, and A.H. Al-Farhan. 2001. Response of Najdi sheep to oral admittance of *Citrullus colocynthis* fruits, *Nerium oleandes* leaves or their mixture. *Small Ruminant Research* 40 (3): 239–44.

Ali, M.M., M.A. Badri, L.M. Hassan, and I.V. Springuel. 1997. Effect of physiogeographical factors on desert vegetation, Wadi Allaqi Biosphere Reserve, Egypt. *Ecologie* 28 (2): 119–28.

Ali, M., M. Badri, S.N. Moalla, and I.D. Pulford. 2001. Cycling of metals in desert soils: effects of *Tamarix nilotica* inundation by lake water. *Environmental Geochemistry and Health* 29–99:1–10.

Badri, M.A. and A.I. Hamed. 2000. Nutrient value of plants in the extremely arid environment (Wadi Allaqi Biosphere Reserve, Egypt). *Journal of Arid Environments* 44:347–56.

Ball, John. 1912. *The Geography and Geology of South-Eastern Egypt.* Cairo: Survey Department.

Bauer, P.F. 1990. A model concept for utilization of *Balanites aegyptiaca* fruits for the production of vegetable oil and animal feed. UNIDO Report US/GLO/84/233, distribution limited.

Belal, A., B. Leith, J. Solway, and I. Springuel. 1998. *Environmental Valuation and Management of Plants in Wadi Allaqi, Egypt, Report.* International Development Research Center (IDRC) Canada, file:95-1005-01/02 127-01.

Belal, A., I. Pulford, H. Marei, S.D. McGregor, H. El-Shaikh, and R. Salem. 1995. Physical and chemical characteristics of fuel wood species in Wadi Allaqi, Lake Nasser area, Egypt. Allaqi Project Working Paper Series No. 29.

Bishai, H.M., S.A. Abdel-Malek, and M.T. Khalid. 2000. *Lake Nasser.* Cairo: National Biodiversity Unit, No.11, Egyptian Environmental Affairs Agency.

Bond, W.J. 1989. Describing and conserving biotic diversity. In *Biotic diversity in southern Africa: concepts and conservation*, ed., B.J. Huntley. Cape Town: Oxford University Press.

Booth, F.E.M., and G.E. Wickens. 1988. Non-timber uses of selected arid zone trees and shrubs in Africa. *FAO Conservation Guide* 19:18–27.

Boulos, L. 1983. Medicinal plants of North Africa. Michigan: Reference Publications.

Briggs, John and Mohamed Badri. 1997. Vegetation uses and ecological values in Wadi Allaqi: a participatory rural appraisal approach. Allaqi Project Working Paper Series No. 30.

Briggs, J. 1989. Human activity in Wadi Allaqi, April, 1989: a preliminary report. Allaqi Project Working Paper Series No. 3.

Briggs, J. 1991. Pastoralism in Wadi Allaqi region, Egypt: An economy under pressure. In *Pastoral economies in Africa and long term responses to drought*, ed., J. Stone, 197–205. Aberdeen: Aberdeen University Press.

Briggs, J. 1995. Environmental resources: Their use and management by the Bedouin of the Nubian Desert of southern Egypt. In *People and environment in Africa*, ed., J.A. Binns, 61–67. Chichester, UK: John Wiley.

Briggs, J. and M. Badri. 1997. Vegetation uses and ecological values in Wadi Allaqi: A participatory rural appraisal approach. Allaqi Project Working Paper Series No. 30.

Briggs, J., M. Badri, and A.M. Mekki. 1999. Indigenous knowledges and vegetation use among Bedouin in the Eastern Desert of Egypt. *Applied Geography* 19:87–103.

Briggs, J.A., G. Dickinson, K.J. Murphy, I.D. Pulford, A.E. Belal, S. Moalla, I. Springuel, S. Ghabbour, and A.M. Mekki. 1993. Sustainable development and resource management in marginal environments: natural resources and their use in the Wadi Allaqi region of Egypt. *Applied Geography* 13:259–84.

Briggs, J., I. Pulford, M. Badri, and A.S. Shaheen. 1998. Indigenous and scientific knowledges: The choice and management of cultivation sites by Bedouin in Upper Egypt. *Soil Use and Management* 14:240–45.

Briggs, J. and J. Sharp. 2004. Indigenous knowledges and development: a postcolonial caution. *Third World Quarterly* 25:661–76.

Briggs, J., J. Sharp, N. Hamed, and H. Yacoub. 2003. Changing gender relations, changing environmental knowledges and livestock management in Upper Egypt. *Geographical Journal* 169:313–25.

Briggs, J., J. Sharp, H. Yacoub, N. Hamed, and A. Roe. 2007. The nature of indigenous knowledge production: evidence from Bedouin communities in southern Egypt. *Journal of International Development*, 19, 239–251.

Cultivation of medicinal plants, annual report. 1995. Submitted to the Ministry of International Co-operation, Assiut University, Egypt.

Cultivation of medicinal plants, mid-year report. 1995. Submitted to the Ministry of International Co-operation, Assiut University, Egypt.

Dafalla, H. 1975. *The Nubian exodus.* London: C. Hurst and company.

Degen, A.A., T. Mishorr, H.P.S. Makkar, M. Kam, R.W. Benjamin, K. Becker, and H.J. Schwartz. 1998. Effects of *Acacia saligna* with and without administration of polyethylene glycol on dietary intake in desert sheep. *Animal Science* 67 (3): 491–98.

Dickinson, G. 1991. Conservation Management Strategies for the Wadi Allaqi Protected Area. Allaqi Project Working Paper Series No. 15.

Drouhin, G. 1970. Alternative uses of arid regions. In *Arid lands in transition,* ed. E. Dregne. Publication no. 90 of the American Association for the Advancment of Science. Washington D.C.: American Association for the Advancement of Science.

Entz, B. 1976. Lake Nasser and Lake Nubia. In *The Nile: biology of an ancient river,* ed. J. Rzóska. The Hague: Junk.

Findlay, A.M. 1998. Policy implications on population growth in arid environments. In *Population and Environment in Arid Regions*, eds. J. Clarke and D. Noin. Paris: UNESCO and Pearl River, NY: Parthenon Publishing.

Kassas, M. 1952. Habitat and plant communities in the Egyptian desert. *Journal of Ecology* 40:342–51.

Kassas, M. 2002. *Environmental change and desert development: New water, new opportunities in southern Egypt. Symposium report, UESD.* Aswan: South Valley University.

Khalil, R. and A. Ali. 2000. *Egypt's Natural Heritage.* Cairo: privately published.

Al-Khodari, Nabil. 2003. *The Nile river: Challenges to sustainable development,* CEO, Nile Basin Society, Canada, Presentation to the River Symposium, http://www.infoplease.com/ce6/world/A0805772.html.

Kumar, R. and J.P.F. D'Mello. 1995. Antinutritional factors in forage legume. In *Tropical Legumes in Animal Nutrition*, eds. J.P.F. D'Mello, and C. Devendra. Wallingford: CAB International.

Giffard, P.L. 1974. *L'Arbre dans le Paysage Senegalais: Sylviculture en Zone Tropicale Seche.* Dakar: Centre Technique Forestier Tropical.

Hall, J.B. 1992. Ecology of a key African multipurpose tree species, *Balanites aegyptiaca* (Balanitaceae): the state-of-knowledge. *Forest Ecology and Management* 50:1–30.

Hamed, A.I. and N.A. El-Emary. 1999. Triterpene saponins from *Glinus lotoides var. dictamnoides. Phytochemistry* 50:477–80.

Hamed, N., H. Ibrahim, A.M. Mekki, I. Springuel, H. Yacoub, J. Briggs, A. Roe, and J. Sharp. 2002. Indigenous environmental knowledges and sustainable development in semi-arid Africa. Report. Department for International Development (DFID). Project Number R7906.

Hamed, A.I., I. Springuel, N.A. El-Emary, H. Mitome, H. Miyaoka, and Y. Yamada. 1996. Triterpenoidal Saponin Glycosides from *Glinus lotoides var. dictamnoides. Phytochemistry* 43 (1): 183–88.

Hamed, A.I., I. Springuel, N.A. El-Emary, H. Mitome, H. Miyaoka, and Y. Yamada. 1997. A Phenolic cinnamate dimmer from *Prosalea plicata*. *Phytochemistry* 43 (1): 183–88.

Healy, R.G. 1997. Sustainable development projects in biosphere reserve: recurring pitfalls and how to avoid them. *Ecodecision* 23:34–38.

Hoath, R. 2003. A field guide to the mammals of Egypt. Cairo: American University in Cairo Press.

Hussein, H.A. 1997. Studies on the nutritive values of some natural plants in Wadi Allaqi Conservation Area in Aswan. PhD Thesis. Department of Agricultural Sciences, Institute of Environmental Studies and Research, Ain Shams University, Cairo.

Latif, A.F.A. 1974. *Fisheries of Lake Nasser.* Aswan: Aswan Regional Planning Lake Nasser Development Center.

Marei, H., R. Salem, A. Long, and A. Belal. 1995. Biomass determination and growth model of *Tamarix nilotica* shrub in Wadi Allaqi, Lake Nasser area, Aswan, Egypt. Allaqi Project Working Paper Series No. 26.

Mekki, Abdel-Moneim and Gordon Dickinson. 1990. Geology, Mineral Resources and Water Conditions in the Wadi Allaqi area Allaqi Project Working Paper Series No. 9.

Mikhail, W.Z.A. 1993. Effect of soil structure on soil fauna in a desert wadi in southern Egypt. *Journal of Arid Environment* 24:321–31.

Miller, Anthony G. and Miranda Morris. 1988. *Plants of Dhofar. The southern region of Oman, traditional, economic and medicinal uses.* Oman: Office of the Adviser for Conservation of the Environment, Royal Court.

Moalla, S.N. and Pulford, I.D. 1993a. Survey of Soil Resources in Wadi Allaqi. Allaqi Project Working Paper Series No. 18.

Moalla, S.N. and Pulford, I.D. 1993b. Processes Influencing Soil Formation and Soil Properties in Wadi Allaqi. Allaqi Project Working Paper Series No. 19.

Moalla, S.N. and I.D. Pulford. 1995. Mobility of metals in Egyptian desert soils subject to inundation by Lake Nasser. *Soil Use and Management* 11:94–98.

Mohamed, A.I., A.M. Mekki, and J. Briggs. 1991. The social and demographic structure of Wadi Allaqi. Allaqi Project Working Paper Series No. 16.

Murphy, K., I.D. Pulford, G. Dickinson, J. Briggs, and I. Springuel. 1992. Ecological resources for conservation and development in Wadi Allaqi, Egypt. *Botanical Journal of the Linnean Society* 108: 131–41.

National Research Council. 1983. *Firewood crops: Shrub and tree species for energy production,* Vol. 2. Washington D.C.: National Academy of Sciences.

Nour, A.A.M., AH.R. Ahmed, and A.G.A. Abdel-Gayoum. 1985. A chemical study of *Balanites aegyptiaca* fruits grown in Sudan. *Journal of the Science of Food and Agriculture* 36:1254–58.

Noy-Meir, I. 1973. Desert ecosystem: environment and producers. *Annual Review of Ecology and Systematics* 4:25–41.

Piotrovsky, B. 1965. The early dynasty settlement of Khor Daoud and Wadi-Allaki. The ancient route of the gold mines. In *Fouilles en Nubie, 127–40. Cairo:* n.p., 1965.

Pulford, I.D. 1989. The Soils of Wadi Allaqi: A General Overview. Allaqi Project Working Paper Series No. 2.

Pulford, I.D., K.J. Murphy, G. Dickinson, J.A. Briggs, and I. Springuel. 1992. Ecological resources for conservation and development in Wadi Allaqi, Egypt. *Botanical Journal of the Linnean Society* 108:131–41.

Radwan, U.A., A.S. Shaheen, and I. Pulford. 1990. Properties of Recently Inundated Soils in Wadi Allaqi. Allaqi Project Working Paper Series No. 8.

Radwan, U.A., T. Radwan, M. Sheded, and I. Pulford. 1997. Soil salinity and requirements for irrigation in Wadi Allaqi. Allaqi Project Working Paper Series No. 31.

Raheja, P.C. 1973. *Lake Nasser in man made lakes: their problems and environmental effects*, eds., W.C. Ackermann, G.F. White, E.B. Worthington. Washington D.C.: American Geophysical Union.

Ramana, D.B.V., S. Singh, K.R. Solanki, and A.S. Negi. 2000. Nutritive evaluation of some nitrogen and non-nitrogen fixing multipurpose tree species. *Animal Feed Science and Technology* 88 (1–2): 103–11.

Said, R. 1993. *The river Nile.* London: Pergamon Press.

El-Sayed, Sayed and Gert L. van Dijken. 1995. The southeastern Mediterranean ecosystem revisited: thirty years after the construction of the Aswan High Dam. *Oceanography*, Texas A&M University. http://ocean.tamu.edu/ Quarterdeck/QD3.1/Elsayed/elsayed.html (22 October 2008).

Selim, Sayed A. Ahmed. 1990. Applications of remote sensing and geographical information systems in the Wadi Allaqi area, south-east of Egypt. Allaqi Project Working Paper Series No. 6.

Seville Strategy, 1995. The Seville Strategy for Biosphere Reserves. http://74.6.239.67/search/cache?ei=UTF8&p=Seville+Strategy+1995&fr= yfpt501&fp_ip=RU&u=www.unesco.org/mab/doc/Strategy.pdf&w=seville +strategy+1995&d=FOxZWfReReR2&icp=1&.intl=us.

Sharp, J., J. Briggs, H. Yacoub, and N. Hamed. 2003. Doing gender and development: Understanding empowerment and local gender relations. *Transactions, Institute of British Geographers* 28 (3): 281–95.

Solway, J. and A.M. Mekki. 1999. Socio-economic system of Wadi Allaqi. Allaqi Project Working Paper Series No. 33.

Springuel, I. 1991. Bases for the economic utilisation and conservation of vegetation in Wadi Allaqi conservation area, Egypt. *African Journal of Agricultural Science* 18 (2): 65–84.

Springuel, Irina. 1994. Basis for economic utilization and conservation of vegetation in Wadi Allaqi conservation area, Egypt. Allaqi Project Working Paper Series No. 13.

Springuel, I. 1997. Vegetation, land use and conservation in the southern eastern desert of Egypt. In *Reviews in ecology: desert conservation and development*, eds. H. Barakat and A. Hegazy, 177–206. Cairo: privately published.

Springuel, I. 2001. Indigenous agroforestry for sustainable development of the area around Lake Nasser, Egypt. In *Sustainable Land Use in Deserts*, eds., S.W. Breckle, M. Veste, and W. Wucherer. Berlin: Springer.

Springuel, I. and O. Ali. 2005. River Nile wetlands: An ecological perspective. In *The World's Largest Wetlands: Ecology and Conservation*, eds., P.A. Keddy and L.H. Fraser, 316–61. Cambridge: Cambridge University Press.

Springuel, I. and A. Belal in cooperation with managers of WABR. 2003. *Report on Wadi Allaqi Biosphere Reserve, ten years periodic review.* Submitted to UNESCO Paris, Aswan. Limited distribution.

Springuel, I., J. Briggs, J. Sharp, H. Yacoub, N. Hamed. 2001. Women, environmental knowledge and livestock production in Wadi Allaqi biosphere reserve. Allaqi Project Working Paper Series No. 37.

Springuel, I., N. El-Emary, and A.I. Hamed. 1993. Medicinal Plants in Wadi Allaqi. Allaqi Project Working Paper Series No. 21.

Springuel, I., A.H. Hussein, M. El-Ashri, M., Badri, and A. Hamed. 2003. Palatability of fodder plants, *Arid Ecosystems* 9 (18): 30–39.

Springuel, I. and A.M. Mekki. 1994. Economic value of desert plants: *Acacia* trees in Wadi Allaqi Biosphere Reserve. *Environmental Conservation* 21 (1): 41–48.

Springuel, I., K. Murphy, and M. Sheded. 1997. The plant biodiversity of the Wadi Allaqi Biosphere Reserve (Egypt): Impact of Lake Nasser on desert wadi ecosystem. *Biodiversity and Conservation* 6:1259–75.

Springuel, I., U.A. Radwan, P.K. Biswas, D. Hileman, G. Huluka, and M. Alemayenu. 1992. Water conditions of the soils in Wadi Allaqi, Lake Nasser Region. In *Proceedings of the First National Conference on: The Future of Land Reclamation and Development in Egypt*, ed., M.A. Kishk. 307–26. Minia: n.p.

Springuel, I., A.S. Shaheen, and K.J. Murphy. 1995, Effects of grazing, water supply, and other environmental factors on natural regeneration of *Acacia raddiana* in an Egyptian Desert Wadi System. In *Rangelands in a sustainable biosphere*, ed., Neil E. West. Proceedings of the Fifth International Rangeland Congress, Vol.1:529–30. Colorado: Society for Range Management.

A status report on the protected area network of Egypt. 2003. Cairo: Egyptian Environmental Affairs Agency.

Sutton, M.D. 1976. Recreation and tourism in arid lands. In *Arid – land ecosystems: structure, functioning and management*, eds., D.W. Goodall and R.A. Perry. Cambridge: Cambridge University Press

Tourism Development Authority, 1999. *Eco-tourism and the Egyptian context.* Cairo: Tourism Development Authority.

de Vrie M. and E. van Zanten. 1993 Report. A hydrogeological survey of the Wadi Allaqi, Egypt with emphasis on the influence of the Aswan High Dam Lake. Amsterdam: Institute of Earth Science, Free University.

White, Gebert F. 1988. The environmental effects of the High Dam at Aswan. *Environment* 30 (7): 4–40.

White, T. 1995. Overgrazing in Wadi Allaqi. Allaqi Project Working Paper Series No. 25.

Index